Latin American Mathematics Series

SBMAC Collection on Applied and Computational Mathematics

The SBMAC Collection on Applied and Computational Mathematics, as part of the Latin American Mathematical Series, publishes significant contributions in the fields of applied and computational mathematics, scientific computing, and related areas. This collection encompasses a range of content from professional to academic, including monographs, advanced textbooks, and contributed volumes. All volumes undergo rigorous peer review by subject experts.

The Brazilian Society of Computational and Applied Mathematics (Sociedade Brasileira de Matemática Aplicada e Computacional – SBMAC) is a professional association focused on computational and industrial applied mathematics. The society actively furthers the development of mathematics and its applications in scientific, technological, and industrial fields. SBMAC has contributed to the advancement of applications of mathematics in science, technology, and industry, encouraged the development and implementation of effective mathematical methods and techniques for the benefit of science and technology, and promoted the exchange of ideas and information across diverse areas of application.

Júlio Araújo • Mitre C. Dourado • Fábio Protti •
Rudini M. Sampaio

Introduction to Graph Convexity

An Algorithmic Approach

Springer

Júlio Araújo
Mathematics Department
Federal University of Ceará
Fortaleza, Ceará, Brazil

Mitre C. Dourado
Institute of Computing
Federal University of Rio de Janeiro
Rio de Janeiro, Rio de Janeiro, Brazil

Fábio Protti
Institute of Computing
Fluminense Federal University
Niterói, Rio de Janeiro, Brazil

Rudini M. Sampaio
Computer Science Department
Federal University of Ceará
Fortaleza, Ceará, Brazil

ISSN 2524-6755 ISSN 2524-6763 (electronic)
Latin American Mathematics Series
SBMAC Collection on Applied and Computational Mathematics
ISBN 978-3-031-84130-9 ISBN 978-3-031-84128-6 (eBook)
https://doi.org/10.1007/978-3-031-84128-6

Mathematics Subject Classification: 05C90, 52A99, 68R10, 05C62, 05C85

English translation of the first original Brazilian Portuguese edition published by Editora do IMPA, Rio de Janeiro, 2023.
The original submitted manuscript has been translated into English. The translation was done using artificial intelligence. A subsequent revision was performed by the author(s) to further refine the work and to ensure that the translation is appropriate concerning content and scientific correctness. It may, however, read stylistically different from a conventional translation.
Translation from the Portuguese language edition: "Uma introdução à convexidade em grafos" by Júlio Araújo et al., © The Authors 2023. Published by Editora do IMPA. All Rights Reserved.

This Springer imprint is published by the registered company Springer Nature Switzerland AG
The registered company address is: Gewerbestrasse 11, 6330 Cham, Switzerland

This book is dedicated to the exceptional professors and researchers Jayme Luiz Szwarcfiter (above) and Nelson Maculan Filho (below), who recently turned 80 years old. We wish them health so that they can continue to be these examples to be followed by the future generations.

Foreword

It is a great pleasure to have the opportunity to write this foreword. Both for the authors and due to the chosen theme. The book comes to fill a gap in the absence of a text of this nature in the literature. The only material that comes close to this book is from an author at the University of Barcelona, but it basically differs from the present proposal. There are no other records of books in the world literature that deal with convexity with this focus, in which aspects of computational complexity play a predominant role.

After introducing some basic notions of convexity, the text presents the protagonist of the book: Graph Convexity. The emphasis is on the area of convexity that contains almost all the work done so far, namely, path convexity in graphs.

Next, convexity parameters are presented, in which most of the research has been concentrated so far. The determination of these parameters, in different convexities, is one of the main motivations of the work done in convexity.

In general, convexity parameters can be grouped into two distinct types. Those originating from the area of abstract convexity, comprising algebraic and geometric studies, and those motivated by the area of computational complexity. The first of these types includes parameters such as Carathéodory number, Radon number, Helly number, and Rank. These parameters have been treated in the literature for several years, at least since the beginning of the last century or before. Meanwhile, parameters motivated by computational complexity are much more recent. Among these are the hull number, convexity number, interval number, among others. The book covers the study of the computation of these two types of parameters.

In the following chapter, concepts of geometric convexity in graphs are described. In this context, several classes of graphs known in the context of computational complexity naturally arise, such as interval graphs, proper interval, cographs, chordal graphs, strongly chordal, among others.

In addition, the text presents, in detail, the study of the determination of parameters in different convexities. The book also contains a description of some applications of convexity, including in games.

I have been following the development of work involving Graph Convexity in Brazil since its inception. I am a witness to the progress that this area has experienced in the country in recent times. In fact, I would venture to say that Brazil is possibly the main center of development of research in Graph Convexity.

As far as I know, the first works in convexity in Brazil were carried out more than 20 years ago. Since that time, until today, the development of the area in the country has been enormous.

The first works were carried out at the Federal University of Rio de Janeiro. Shortly after, they reached the Federal Fluminense University, the Federal University of Ceará, the Federal University of Goiás, and the Federal University of Minas Gerais. Coincidentally or not, the universities of origin of the authors of this work are exactly the first three.

In addition, shortly after the first works in the area were carried out, through a CAPES-DAAD cooperation project, UFRJ received a visit from researcher Dieter Rautenbach, at the time linked to the University of Ilmenau, who later transferred to the University of Ulm, where he is currently located. This cooperation continues to this day, including the inclusion of researcher Lúcia Penso, also from the University of Ulm. Dieter's contributions are numerous, in the works carried out, in convexity and related topics, with researchers from the country. In fact, this cooperation was one of the most relevant factors for the development of the area of convexity in Brazil. To illustrate the richness of themes and the breadth of the area, when a group of researchers gathered around a table to discuss what research topics would be considered in that meeting, it was common to hear the following dialogue:

What will we discuss today?
And about which convexity?
And about which parameter?
And with what objective?

In reality, the possibilities for research on the topic are diverse. The dialogue well illustrates this fact. To conclude, I would like to once again emphasize the excellence, quality, and timeliness of the authors' work.

Rio de Janeiro
June 2023

Jayme L. Szwarcfiter

Preface

As we will see, Graph Convexity is a young research area, which has only recently begun to be studied, but already has a vast literature of scientific papers. Because of this youthfulness, unlike older areas of Mathematics, there are good possibilities for researchers and students to produce relevant contributions. Much of the recent scientific research in Graph Convexity was published in the last 10 years. Therefore, this book is an invitation for you to get excited about this beautiful area and help our community to develop it even more.

Graph Convexity was born as a work of transposing the concepts of Convex Geometry to the field of Combinatorics. Thus, a geodesic connecting two points in a space naturally translates into a minimum path between two vertices of a graph; the convex hull of a set of points has its combinatorial analogue determined by the iterative operation of adding new vertices that are found in minimum paths between vertices already included in the hull. Such concepts form the essence of the so-called Geodesic Convexity, from which other convexities were defined, sometimes considering other types of paths besides the minimum paths (such as, for example, the induced paths associated with Monophonic Convexity), sometimes considering other rules for the determination of convex hulls.

A particular characteristic of Graph Convexity in Brazil is its emergence within the area of Computing, specifically in the subarea of Algorithms and Complexity. The group of Brazilian researchers who initially dedicated themselves to research in Convexity soon considered the problems pertinent to the topic from an algorithmic point of view. For example: What is the complexity of the problem of determining the smallest set of vertices whose convex hull is equal to the set of vertices of the graph? Being this problem NP-hard, for which classes of graphs can we design polynomial time algorithms that solve this problem? And several other questions with the same *algorithmic appeal* have been posed since the origin of research in our country. This does not mean that this book is aimed at the Algorithms and Complexity community, as it introduces the topics always initially focusing on their theoretical foundation. We believe that mathematicians and students from other areas will undoubtedly benefit from the content of this text.

How to Use This Book

This book was written as part of a minicourse of the 34° Brazilian Mathematics Colloquium (IMPA) in July 2023. There has long been a desire to write a book in this area by some researchers and this was an excellent opportunity for that. At this point, we have much to thank the IMPA team, especially Prof. Paulo Ney de Souza, who helped us with many suggestions and corrections, as well as very relevant cultural information.

It is aimed at advanced undergraduate and postgraduate students in mathematics or computer science. We sought to provide self-contained content, so that students can understand the topic without the need to resort to other sources. However, in some chapters, especially on more technical topics, there are only mentions of certain results, with the appropriate bibliographic reference. The book is divided into two parts. Part I contains the definitions and general results of the area of Graph Convexity, while Part II contains more specific definitions and results, such as more technical results of the main convexities and some recent applications, such as convexity games. Chapters 1 and 2 are strongly recommended for reading as they contain the main definitions, while Chap. 3 contains many general examples for easy understanding.

Fortaleza, Brazil — Júlio Araújo
Rio de Janeiro, Brazil — Mitre C. Dourado
Niterói, Brazil — Fábio Protti
Fortaleza, Brazil — Rudini M. Sampaio
January 2024

Acknowledgments

The authors would like to initially thank Prof. Celina de Figueiredo for the essential encouragement given for the realization of this work. We also thank Emanuel Elias for the great help with the bibliographic review.

JA, MCD, FP, RMS

I thank Prof. Jayme Szwarcfiter for introducing me to the area of Graph Convexity in 2010, in Grenoble, at an event we both attended. Since then I have been fascinated by the area and have been continuously working on this theme. I dedicate this work to my wife Karol Maia and our son Igor.

Júlio Araújo

I would like to thank Prof. Jayme Luiz Szwarcfiter for having predicted that the study of Computational Complexity of Graph Convexity would be fruitful and, mainly, for having told us in advance. Will the area continue to grow? Only he can answer.

Mitre C. Dourado

My thanks go to Prof. Jayme Luiz Szwarcfiter, without whom nothing I accomplished as a researcher would have materialized. I thank God for having placed brilliant and competent people by my side, as they encouraged me to evolve professionally. I dedicate this book to my wife Érica and our children Olívia, Inês, and Eduardo.

Fábio Protti

I thank Jayme for introducing me to this topic in Fortaleza in 2011, giving me the opportunity to be his co-author. I dedicate this book to Ana Karina, my girlfriend since I was 16, and to our children Samuel, Catarina, and Miguel, who were patient during the nights spent writing this book. May God always bless them and all those who read these pages!

Rudini M. Sampaio

Contents

Part I
Fundamentals of Convexity in Graphs

Chapter 1
Basic Concepts of Convexity

Convexity is a classic theme, studied in many different areas of mathematics. Among the oldest contributions, there are several in the book "Elements," written by Euclid around 300 BC, as well as the first precise definition of a convex curve or surface, given by Archimedes around 250 BC. Ancient mathematicians and physicists, such as Kepler, Descartes, Euler, Fourier, Gauss, and Cauchy, made important contributions to the field of convexity. We recommend the chapter "History of Convexity" from the book "Handbook of Convex Geometry" by Gruber and Wills (1993) for more details. Other important references are the book "Theory of Convex Structures" by Van de Vel (1993) and the recent book "Combinatorial Convexity" by Bárány (2021).

Despite these ancient contributions, the field of convexity in geometry (or convex geometry) only became an independent branch of mathematics in the early twentieth century, mainly with the systematic development of the theory of convexity by Hermann Minkowski (1903, 1911), who among various contributions developed the Minkowski four-dimensional spacetime, which was of fundamental importance for Einstein's theory of relativity. Afterwards, important results enriched the field, such as the theorems of Carathéodory (1911), Radon (1921), Helly (1923), and Erdős and Szekeres (1935), which laid the foundations for the area of combinatorial convexity.

The study of convexity applied to graphs began much later, in the 1970s. Paul Erdős, one of the most prolific mathematicians in history, published in 1972 with other authors one of the first papers on graph convexity, focused on tournaments.[1] According to Duchet (1987), the first paper explicitly on convexity in general graphs in English is the paper "Convexity in graphs" by Harary and Nieminem (1981), in which the iteration time was introduced, one of the convexity parameters studied in this book. One of its authors, Frank Harary, recognized as one of the fathers of

[1] Tournaments are directed graphs obtained from the orientation of the edges of a complete graph.

J. Araújo et al., *Introduction to Graph Convexity*, SBMAC Collection on Applied and Computational Mathematics, https://doi.org/10.1007/978-3-031-84128-6_1

modern graph theory, introduced some combinatorial games (Harary 1984) related to graph convexity, which will be seen in the last chapter.

In the 1980s, the research on graph convexity grew considerably and developed with the contribution of several researchers, notably Robert E. Jamison, who published many papers on the subject, which are cited to this day, such as the papers (Jamison 1981, 1982), in which the rank of a graph and the monophonic convexity in graphs are introduced.[2]

At this point, you may wonder about the relevance of studying convexity in graphs, disassociated from geometry. In one of his first papers, "A perspective on abstract convexity," still on old typewriters, Jamison (1982) asks: "*Why abstract convexity?*" Before providing several strong arguments, which are beyond the scope of this introduction, he says the following:

> One possible reason to abstract convexity is that it can be done so easily. It is not at all hard to invent a reasonable set of axioms for a system of convex sets, a convex hull operator, or a generalized interval function. But this can hardly be considered a satisfactory answer. (Jamison 1982)

In the last 15 years, the area of graph convexity has seen a resurgence as various researchers of theoretical computer science began studying graph convexity parameters from a computational complexity perspective, making this topic quite active in recent research, including applications of diffusion and influence maximization in social networks (Sect. 9.1), among others. Much of this is due to the role of the brilliant mathematician Jayme Szwarcfiter who greatly stimulated research in this area.

But, before delving into the main topic of graph convexity, we will provide a brief introduction about the area of convex geometry. This will be important, as several concepts used in graph convexities are the same as in convex geometry, such as convex hull; extreme points; Carathéodory, Radon, and Helly numbers; general position; and convex position.

1.1 Convex Geometry: Basic Concepts

Let $d \geq 2$ be an integer and $\mathbb{R}^d$ the d-dimensional real space. In geometry, a set $S \subseteq \mathbb{R}^d$ is said to be *convex* if S contains any segment between two points of S. As an example, circles are convex in $\mathbb{R}^2$ and spheres are convex in $\mathbb{R}^3$, but circumferences and spherical shells are not convex. The *convex hull* of a set $S \subseteq \mathbb{R}^d$, denoted by $\text{conv}(S)$, is defined as the smallest convex set that contains S. Figure 1.1 shows a non-convex set of the plane $\mathbb{R}^2$ and its convex hull.

The following theorem shows a very important property of convex sets and its proof will be left for Exercise 1.1.

---

[2] The rank is one of the main parameters of graph convexity and the monophonic convexity is one of the main graph convexities. Both are studied in this book.

Fig. 1.1 Example of convex hull in the Euclidean plane

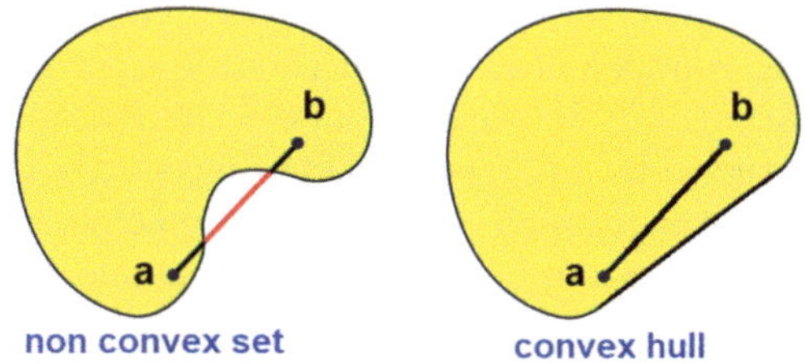

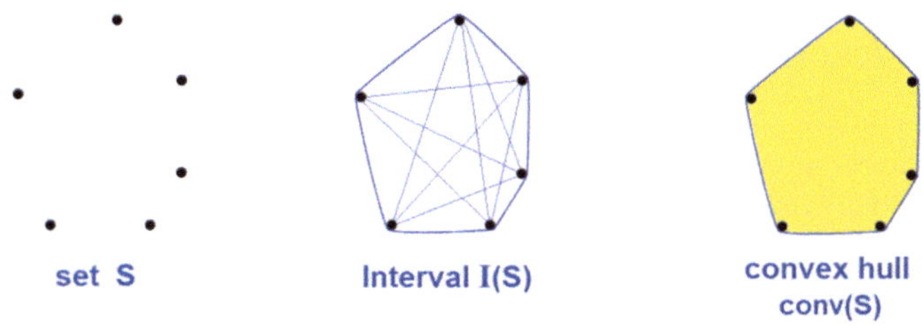

Fig. 1.2 Convex set in yellow is the convex hull of its extreme points

Theorem 1.1 *Let $d \geq 2$ be an integer. The intersection of two convex sets in $\mathbb{R}^d$ is a convex set. Furthermore, $\emptyset$ and $\mathbb{R}^d$ are convex.*

A natural problem is to obtain the smallest subset S' of a given set $S \subseteq \mathbb{R}^d$ such that $\text{conv}(S') = \text{conv}(S)$. The answer is given by the following theorem, whose first version is due to Minkowski (1911): S' contains the extreme points of $\text{conv}(S)$. To state it, we need some basic definitions.

Let $S \subseteq \mathbb{R}^d$. We say that S is *open* if every point $x \in S$ has a neighborhood in S, i.e., there exists $r > 0$ such that every point at distance r from x also belongs to S. We say that S is *closed* if its complement is open. We say that S is *bounded* if there exists $r > 0$ such that the distance between any two points of S is less than or equal to r. Finally, if S is convex, we say that $x \in S$ is an *extreme point* of S if $x \notin \text{conv}(S \setminus \{x\})$, i.e., x does not belong to any segment between two other points of S. We denote by $\text{Ext}(S)$ the set of extreme points of S.

Theorem 1.2 (Minkowski–Krein–Milman Theorem) *Every convex subset $S \subseteq \mathbb{R}^d$ that is closed and bounded is the convex hull of its extreme points.*

Minkowski (1911) proved a restricted version of this theorem for $\mathbb{R}^3$, which was extended by Steinitz (1916) to any $\mathbb{R}^d$ with $d \geq 3$. Krein and Milman (1940) proved a generalization of this theorem for non-Euclidean spaces. As an example, Fig. 1.2 shows a convex set in yellow. Observe that it is equal to the convex hull of its extreme points.

We now define some general concepts that are more used in graph convexity, but also have motivation in convex geometry. Given a set $S \subseteq \mathbb{R}^d$, the *interval* of S, denoted by $\text{I}(S)$, is the set of points in some segment between two points of S inclusive. We call $\text{I}(\cdot)$ the *interval function*. See Fig. 1.2 for an example of a set

S, its interval $\mathrm{I}(S)$, and its convex hull $\mathrm{conv}(S)$. Note that it is possible to obtain the convex hull of S by successively applying the interval function. In the example of Fig. 1.2, we see that $\mathrm{conv}(S) = \mathrm{I}(\mathrm{I}(S))$. The *iteration time* of a set $S \subseteq \mathbb{R}^d$, denoted by $\mathrm{ti}(S)$, is defined as the smallest integer k such that $\mathrm{I}^k(S) = \mathrm{conv}(S)$, i.e., k applications of the interval function are needed to obtain the convex hull. In the example of Fig. 1.2, the iteration time $\mathrm{ti}(S) = 2$.

1.2 Classics: Carathéodory, Radon, and Helly

The three classic theorems of Carathéodory, Radon, and Helly "*form the cornerstones of the vast area of Combinatorial Convexity*" (Gruber and Wills 1993) and will be important in the area of graph convexity, leading to the definition of three related parameters: the Carathéodory number, the Radon number, and the Helly number.

Theorem 1.3 (Carathéodory 1911) *Given $S \subseteq \mathbb{R}^d$, if $x \in \mathrm{conv}(S)$, then there exists $C \subseteq S$ with $|C| \leq d + 1$ such that $x \in \mathrm{conv}(C)$.*

Figure 1.3 illustrates Carathéodory's theorem: $x \in \mathrm{conv}(S)$, where S consists of points 1 to 8. Note that x is also in the convex hull of only 3 points of S. In this example from the plane $\mathbb{R}^2$, we have $d = 2$, and therefore, $d + 1 = 3$ points are sufficient.

The *Carathéodory number* $\mathrm{cth}(S)$ of a convex set $S \subseteq \mathbb{R}^d$ is defined as the smallest integer $r \geq 0$ such that, for every $S' \subseteq S$ and $x \in \mathrm{conv}(S')$, there exists $C \subseteq S'$ with $|C| \leq r$ such that $x \in \mathrm{conv}(C)$. With this, Carathéodory's theorem implies that $\mathrm{cth}(S) \leq d + 1$ for every convex set $S \subseteq \mathbb{R}^d$.

According to Van de Vel (1993), it is not difficult to prove (Exercise 1.8) that the Carathéodory number of S is also the size of the largest Carathéodory-independent subset $C \subseteq S$, where we say that C is *Carathéodory independent* if

$$\mathrm{conv}(C) \neq \bigcup_{x \in C} \mathrm{conv}(C \setminus \{x\}).$$

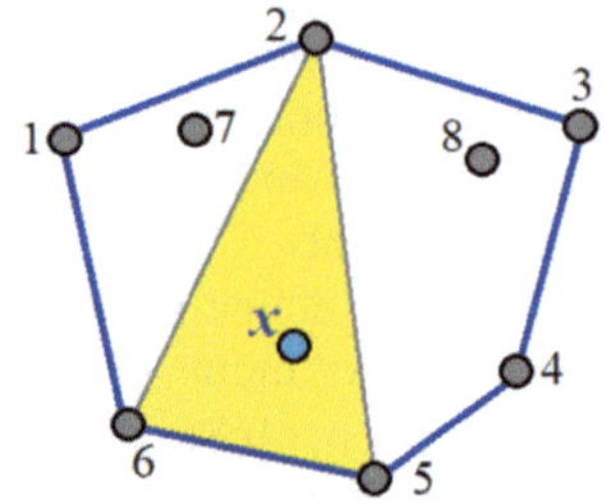

Fig. 1.3 Example of Carathéodory's theorem

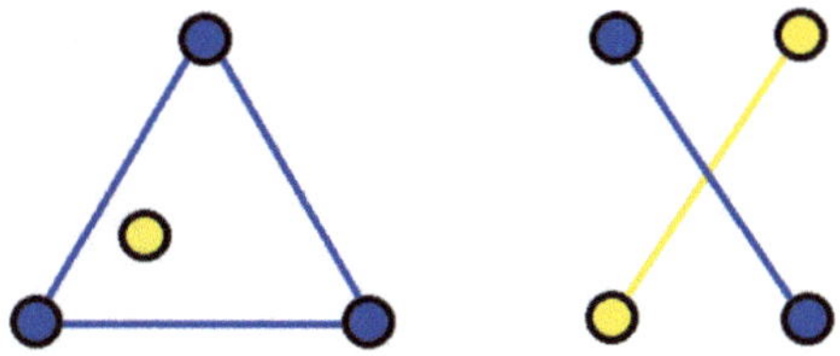

Fig. 1.4 Example of Radon's theorem

We say that a set $S \subseteq \mathbb{R}^d$ is *Radon dependent* if S can be partitioned into two sets whose convex hulls intersect; otherwise, it is *Radon independent.*

Theorem 1.4 (Radon 1921) *Every set $S \subseteq \mathbb{R}^d$ with $d + 2$ points is Radon dependent. That is, every Radon-independent set has at most $d + 1$ points.*

In Fig. 1.4, we have two examples of a set $S \subseteq \mathbb{R}^2$ with 4 points and their bipartitions (in yellow and blue) whose convex hulls intersect. In this example from the plane $\mathbb{R}^2$, we have $d = 2$ and therefore at least $d + 2 = 4$ points are needed. Radon's theorem was generalized for partitions into any number $k > 2$ of subsets (Tverberg 1966).

The *Radon number* $\mathrm{rd}(S)$ of a convex set $S \subseteq \mathbb{R}^d$ is defined as the size of the largest Radon-independent subset $S' \subseteq S$. With this, Radon's theorem implies that $\mathrm{rd}(S) \leq d + 1$ for every convex set $S \subseteq \mathbb{R}^d$.

Theorem 1.5 (Helly 1923) *If $\mathcal{F}$ is a family of convex sets in $\mathbb{R}^d$ such that every $d + 1$ members of $\mathcal{F}$ have a common point, then all members of $\mathcal{F}$ have a common point.*

Helly's theorem, proved in 1913 but only published in 1923, has many generalizations, such as a colored version (Bárány 2021). A typical *Helly-type* theorem has the following format: *if every group of n members of a family of objects has a certain property, then the whole family has that property*. In Theorem 1.5, the *property* is to have a common point.

Figure 1.5 shows an example of Helly's theorem for $\mathbb{R}^2$. Figure 1.6a shows the need for the sets to be convex (in the figure, only one of them is not): the intersection of each 3 sets is non-empty, but the intersection of the 4 sets is empty. Remember that, in $\mathbb{R}^2$, $d = 2$ and, therefore, the intersection of each $d + 1 = 3$ sets should be sufficient. Figure 1.6b shows the need for all intersections of $d + 1$ sets to be non-empty: only one intersection of 3 sets is empty and, therefore, the conclusion of Helly's theorem fails.

The *Helly number* $\mathrm{h}\ell(S)$ of a convex set $S \subseteq \mathbb{R}^d$ with more than one point is the smallest integer $h \geq 2$ such that every family $\mathcal{F}$ of convex subsets of S, in which each h members of $\mathcal{F}$ have a common point, satisfies the property that all members of $\mathcal{F}$ have a common point.

With this, Helly's theorem implies that $\mathrm{h}\ell(S) \leq d + 1$ for every $S \subseteq \mathbb{R}^d$ convex. For example, if $S \subseteq \mathbb{R}^d$ consists of collinear points, then $\mathrm{h}\ell(S) = 2$; if $S \subseteq \mathbb{R}^d$ consists of coplanar points, then $\mathrm{h}\ell(S) \leq 3$.

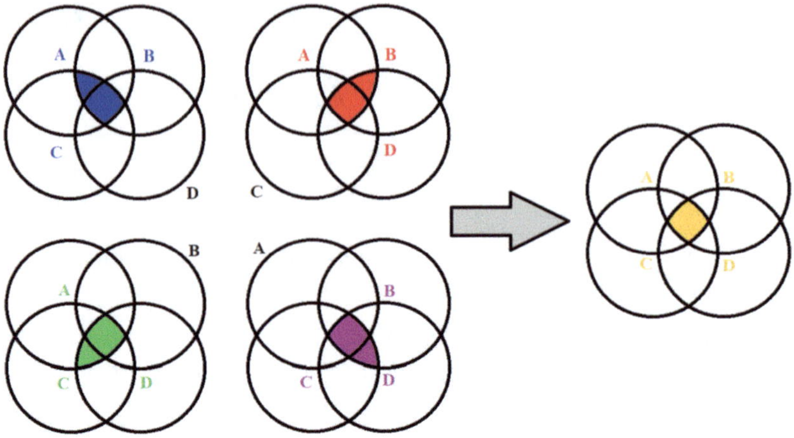

Fig. 1.5 Example of Helly's theorem for $\mathbb{R}^2$: non-empty intersection 3 by 3 implies non-empty intersection of all

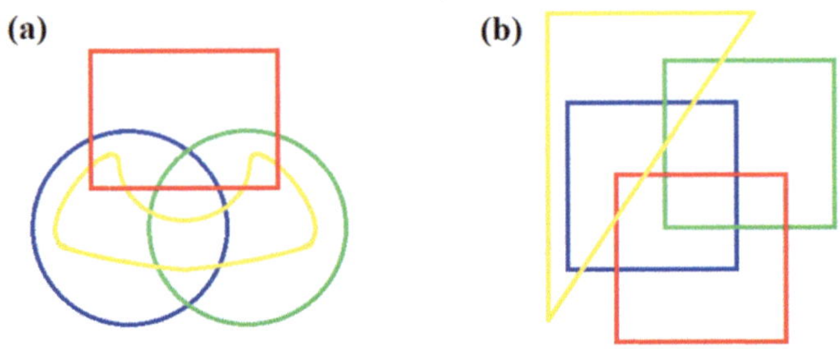

Fig. 1.6 Wrong instances of Helly's theorem

According to Van de Vel (1993), it was independently proved by Calder (1971), Berge and Duchet (1975), and Sierksma (1975) that the Helly number of S is also the size of the largest Helly-independent subset $H \subseteq S$, where we say that H is *Helly independent* if

$$\bigcap_{x \in H} \operatorname{conv}(H \setminus \{x\}) = \emptyset.$$

For example, every maximum Helly-independent set of $\mathbb{R}^1$, $\mathbb{R}^2$, and $\mathbb{R}^3$ is formed by 2 distinct points (segment), 3 noncollinear points (triangle), and 4 noncoplanar points (tetrahedron), respectively.

1.3 General Position and Convex Position

Let $S \subseteq \mathbb{R}^d$ be finite. We say that S is in *general position* if S does not contain 3 collinear points. The *general position number* of S, denoted by $\mathrm{gp}(S)$, is defined as the largest subset of S in general position. When $S = \{1, \ldots, m\}^2 \subseteq \mathbb{N}^2$, this problem becomes the classic *No-Three-in-Line problem* of Dudeney (1917).[3] which remains open for $m > 46$ (Flammenkamp 1998).

We say that S is in *convex position* (or that it is *convexly independent*) if S is the set of extreme points of $\mathrm{conv}(S)$, i.e., no $x \in S$ belongs to $\mathrm{conv}(S \setminus \{x\})$. The famous theorem below is one of the pioneers of Ramsey theory: we can find subsets in convex position of any size in sufficiently large sets in general position.

Theorem 1.6 (Erdős and Szekeres 1935) *Let $d \geq 2$. For every n, there exists N such that every set $S \subseteq \mathbb{R}^d$ in general position with $|S| \geq N$ has a subset $S' \subseteq S$ in convex position with $|S'| = n$.*

This result is a generalization of the *Happy Ending Theorem*, proved by Esther Klein (1933): every set of 5 points in general position in the plane has a subset of 4 points in convex position (convex quadrilateral). Erdős gave it this gracious name (*Happy Ending Theorem*) because, through this research, Esther and George Szekeres became closer, later getting married and living happily ever after: almost 70 years of marriage that ended in 2005 with the death of both in the same day, with only 1 hour of difference, at the age of 95.

With this, we define $ES_d(n)$ as the smallest integer N such that every set $S \subseteq \mathbb{R}^d$ in general position with $|S| \geq N$ has a subset $S' \subseteq S$ in convex position with $|S'| = n$. For example, $ES_2(3) = 3$. The *Happy Ending Theorem* proves that $ES_2(4) = 5$. It is also known that $ES_2(5) = 9$ and $ES_2(6) = 17$, the latter being a computer-assisted proof of Szekeres and Peters (2006). The value of $ES_2(n)$ for $n \geq 7$ is still unknown. In addition, it is known that $ES_d(d+2) = d+3$ and that $ES_{d+1}(n) \leq ES_d(n)$ for all $n, d \geq 2$.

We conclude with a parameter introduced by Jamison (1981) and recently used in graph convexity. The *rank* of a finite set $S \subseteq \mathbb{R}^d$, denoted by $\mathrm{rk}(S)$, is defined as the largest convexly independent subset of S. That is, Erdős–Szekeres theorem states exactly that, for every n, there exists N such that $\mathrm{rk}(S) \geq n$ for every finite set $S \subseteq \mathbb{R}^d$ in general position with $|S| \geq N$.

Exercises

Exercise 1.1 Prove Theorem 1.1. Also show that it does not work for union or difference of sets.

Exercise 1.2 Prove that every $S \subseteq \mathbb{R}^d$ contains the extreme points of $\mathrm{conv}(S)$.

[3] Initially defined as *Puzzle with Pawns* in the book by Dudeney (1917), presented by the prestigious Nature journal in the article Amusements in Mathematics (1917).

Exercise 1.3 Determine the sets of $\mathbb{R}^d$ with iteration time 0, 1, 2, and 3.

Exercise 1.4 Determine the sets of $\mathbb{R}^d$ with Carathéodory number 0, 1, and 2.

Exercise 1.5 Determine the sets of $\mathbb{R}^d$ with Radon number 0, 1, 2, and 3.

Exercise 1.6 Determine the sets of $\mathbb{R}^d$ with Helly number 2, 3, and 4.

Exercise 1.7 Determine the sets of $\mathbb{R}^d$ with rank 2 and 3.

Exercise 1.8 Prove that both definitions of Carathéodory number are equivalent.

Exercise 1.9 Prove the *Happy Ending Theorem*: every set of 5 points in general position in the plane has a subset of 4 points in convex position.

Chapter 2
Convexity in Graphs

2.1 Abstract Convexities on Finite Sets

Theorem 1.1 presents essential properties of convex sets in Euclidean spaces. Such properties are used to define abstract convexities in other topologies. See Van de Vel (1993) and Pelayo (2013). A *convexity* C on a finite set $V \neq \emptyset$ is a family of subsets of V such that $\emptyset, V \in C$ and C is closed under intersection[1] (Levi 1951). That is, $S_1 \cap S_2 \in C$ if $S_1, S_2 \in C$. The members of C are called *convex sets* of C. As an example, there are four possible convexities on $V = \{a, b\}$.

- $C_1 = \Big\{ \emptyset,\ \{a, b\} \Big\}$; $\quad C_3 = \Big\{ \emptyset,\ \{a, b\},\ \{a\} \Big\}$;
- $C_2 = \Big\{ \emptyset,\ \{a, b\},\ \{a\},\ \{b\} \Big\}$; $\quad C_4 = \Big\{ \emptyset,\ \{a, b\},\ \{b\} \Big\}$.

On the other hand, not every family of subsets is a convexity. For example, $\{\emptyset, \{a, b\}, \{a, c\}, \{a, b, c\}\}$ is not a convexity on $V = \{a, b, c\}$, since the intersection of $\{a, b\}$ and $\{a, c\}$ does not belong to the family.

With this broad definition of convexity, we define convex hull and extreme points as done in Chap. 1. The *convex hull* of $S \subseteq V$ in a convexity C on V is the smallest convex set $\text{conv}_C(S)$ containing S. It is easy to check that $\text{conv}_C(\cdot)$ is indeed a *closure operator* (see Exercise 2.1), i.e., for all $S, S' \subseteq V$:

(i)	$S \subseteq \text{conv}_C(S)$	(extensive law).
(ii)	$S \subseteq S' \Rightarrow \text{conv}_C(S) \subseteq \text{conv}_C(S')$	(monotone law).
(iii)	$\text{conv}_C(\emptyset) = \emptyset$	(normalization law[2]).
(iv)	$\text{conv}_C(\text{conv}_C(S)) = \text{conv}_C(S)$	(idempotent law).

[1] For infinite sets, there is one more condition: the infinite union of nested convex sets is convex.

[2] According to Van de Vel (1993), some authors do not consider the property of normalization in the definition of closure operator. We follow the definition of Van de Vel (1993).

J. Araújo et al., *Introduction to Graph Convexity*, SBMAC Collection on Applied and Computational Mathematics, https://doi.org/10.1007/978-3-031-84128-6_2

If $S \subseteq V$ is convex in C, we say that $x \in S$ is an *extreme point* (or *extreme vertex*) of S if $x \notin \text{conv}_C(S \setminus \{x\})$. We denote by $\text{Ext}_C(S)$ the set of extreme points of S. With these definitions of convex hull and extreme points, it is easy to directly define the concepts of convexly independent set, rank, Carathéodory, Radon, and Helly numbers in a convexity C, analogous to the concepts of Chap. 1 (see Exercise 2.2).

We say that a convexity C on V is a *convex geometry*[3] if it satisfies the *Minkowski–Krein–Milman property*: every convex set is the convex hull of its extreme points. In the previous example with $V = \{a, b\}$, only the convexity $C_1 = \{\emptyset, \{a, b\}\}$ is not a convex geometry (or is not geometric) (see Exercise 2.3).

Finally, to be able to define iteration time and general position number, as done in Chap. 1, we need the concept of *interval function*. According to Calder (1971), given a function $\text{I} : \binom{V}{2} \to 2^V$ such that $a, b \in \text{I}(\{a, b\})$, let C be the family of sets S such that $\text{I}(S) = S$, where $\text{I}(S) = \bigcup_{a,b \in S} \text{I}(\{a, b\})$. It is easy to prove that C is a convexity on V (see proof of Lemma 2.1). In the next section, we will see that the main graph convexities are indeed convexities with interval functions of this type, such as the geodesic and the monophonic, which are convexities based on paths in the graph and, thus, determined by pairs of vertices. However, some convexities with broader notions of interval have been recently studied.[4] We say, then, that a function $\text{I} : 2^V \to 2^V$ is an *interval function* on V if for all $S, S' \subseteq V$:

(i) $S \subseteq \text{I}(S)$ (extensive law).
(ii) $S \subseteq S' \Rightarrow \text{I}(S) \subseteq \text{I}(S')$ (monotone law).
(iii) $\text{I}(\emptyset) = \emptyset$ (normalization law).

Note that it is not required to satisfy the idempotent law (iv) of convex hull. We say that an interval function $\text{I}(\cdot)$ on V *induces* the family of sets $S \subseteq V$ such that $\text{I}(S) = S$. Note that this definition of interval function generalizes that of Calder (1971). Below we prove that this family induced by $\text{I}(\cdot)$ is actually a convexity on V.

Lemma 2.1 *Let V be a finite set and $\text{I}(\cdot)$ be an interval function on V. Then the family C of subsets of V induced by $\text{I}(\cdot)$ is a convexity.*

Proof First, note that $\emptyset \in C$ from the normalization law (iii). Moreover, $V \subseteq \text{I}(V) \subseteq V$ from (i) and, therefore, $\text{I}(V) = V$ is in C. Finally, let S_1 and S_2 be two sets in C. By definition, $\text{I}(S_1) = S_1$ and $\text{I}(S_2) = S_2$. From (ii), $\text{I}(S_1 \cap S_2) \subseteq \text{I}(S_1) = S_1$ and $\text{I}(S_1 \cap S_2) \subseteq \text{I}(S_2) = S_2$. Therefore, $\text{I}(S_1 \cap S_2) \subseteq S_1 \cap S_2$. From (i), $S_1 \cap S_2 \subseteq \text{I}(S_1 \cap S_2)$. Therefore, $\text{I}(S_1 \cap S_2) = S_1 \cap S_2$ and consequently $S_1 \cap S_2$ is in C. As C is closed under intersection and $\emptyset, V \in C$, then C is a convexity. □

That is, every interval function induces a unique convexity. Moreover, every convexity can be induced by one (or more than one) interval function, with the

[3] It is also said that C is an *antimatroid* or is a *geometric convexity*. Do not confuse with the area of convex geometry in mathematics, briefly summarized in Chap. 1.

[4] See, for example, Steiner convexity (Cáceres et al. 2008) and r-interval convexities (Dourado et al. 2013a).

convex hull itself being a possible interval function, for example. When in the definition of a convexity C it is explicitly given an interval function $\mathrm{I}_C(\cdot)$ that induces C, we say that C is an *interval convexity*. That is, an interval convexity C on V is defined by explicitly stating which interval function $\mathrm{I}_C(\cdot)$ on V will be considered. The main convexities studied are interval convexities, i.e., the interval function $\mathrm{I}_C(\cdot)$ associated with each of them is clearly defined. Formally, according to Van de Vel (1993), a convexity space is given by a pair (V, C), where C is a convexity on V, while an interval convexity space is given by a pair (V, I_C), with I_C being an interval function on V, which, according to Lemma 2.1, induces a convexity C. We will avoid this terminology and consider for the rest of the book that every convexity C is an interval convexity and, therefore, has an associated interval function I_C.

As in Chap. 1, the convex hull of a set S can be obtained by successively applying the interval function $\mathrm{I}_C(\cdot)$ until obtaining a convex set, that is (Exercise 2.4):

$$\mathrm{conv}_C(S) = \bigcup_{k=1}^{\infty} \mathrm{I}_C^k(S),$$

where $\mathrm{I}_C^k(S)$ is the kth iteration of $\mathrm{I}_C(\cdot)$ on S, i.e., $\mathrm{I}_C^0(S) = S$ and $\mathrm{I}_C^{k+1}(S) = \mathrm{I}_C(\mathrm{I}_C^k(S))$ for all $k \geq 0$. Thus, it is common to say that S *generates* (or *activates*, or *infects*) $\mathrm{I}_C(S)$ in 1 step or $\mathrm{conv}_C(S)$ in certain number of steps in the convexity C. As an example of a strange interval function, see Exercise 2.5.

We can then define for a convexity C on V the counterparts of the parameters of Chap. 1 that depend on the notion of interval, namely, the iteration time and the general position number. The *iteration time* of $S \subseteq V$ is the smallest k such that $\mathrm{I}_C^k(S) = \mathrm{conv}_C(S)$, i.e., k applications of the interval function are needed to obtain the convex hull. In addition, the *general position number* $\mathrm{gp}_C(S)$ is the size of the largest subset of S in general position on C, where S is in *general position* on C if S does not contain 3 elements x, y, z such that $z \in \mathrm{I}_C(\{x, y\})$.

2.2 Graph Convexities

A *graph convexity* C on a finite graph G is an interval convexity on $V(G)$. With this, all concepts of Chap. 1 apply to any graph convexity: convex hull; extreme points; convex geometry; rank; Carathéodory, Radon, and Helly numbers, iteration time; and general position number. Note that all the numerical parameters listed above apply to subsets $S \subseteq V(G)$. We can also define them for the entire graph G, taking $S = V(G)$ (except the iteration time). With this, we have the following 5 parameters for any convexity C on a graph G: $\mathrm{cth}_C(G)$, $\mathrm{rd}_C(G)$, $\mathrm{h}\ell_C(G)$, $\mathrm{rk}_C(G)$, and $\mathrm{gp}_C(G)$. Also let $\mathrm{Ext}_C(G) = \mathrm{Ext}_C(V(G))$: the set of extreme points of $V(G)$ in C.

To complete the list of the most studied convexity parameters, there are 5 that are defined directly on the graph G and not on the set $S = V(G)$ (like the previous ones). The *convexity number* $\text{con}_C(G)$ is the size of the largest convex set of G in C different from $V(G)$. The *hull number* $\text{hn}_C(G)$ is the size of the smallest hull set of G in C, where we say that $S \subseteq V(G)$ is a *hull set* of G in C if $\text{conv}_C(S) = V(G)$. The *interval number* $\text{in}_C(G)$ is the size of the smallest interval set of G in C, where we say that S is an *interval set* of G in C if $I_C(S) = V(G)$. The *iteration time* $\text{ti}_C(G)$ is the largest value $\text{ti}_C(S)$ among all sets $S \subseteq V(G)$. The *percolation time* $\text{tp}_C(G)$ is the largest value $\text{ti}_C(S)$ among all sets $S \subseteq V(G)$ that are hull sets of G in C.

Below is the list of the ten main graph convexity parameters, which are studied in the next chapter. Sometimes we omit the subscript C when it is not relevant.

1. $\text{hn}(G)$: hull number
2. $\text{in}(G)$: interval number
3. $\text{con}(G)$: convexity number
4. $\text{cth}(G)$: Carathéodory number
5. $\text{rd}(G)$: Radon number
6. $\text{h}\ell(G)$: Helly number
7. $\text{rk}(G)$: rank
8. $\text{gp}(G)$: general position number
9. $\text{ti}(G)$: iteration time
10. $\text{tp}(G)$: percolation time

Recall that all these ten parameters depend on the convexity C being considered and that all have been (and still are) the subject of recent scientific research in graph convexity. Bibliographic references will be given in the appropriate sections of each parameter in the following chapters. Once the parameters to be studied in this book were defined, we will now focus on the most studied graph convexities.

2.3 Path Convexities in Graphs

A usual way to define a *convexity* on a graph G is by fixing a family $\mathcal{P}$ of paths in G and obtaining the convexity $C_{\mathcal{P}}$ from the function $I_{\mathcal{P}}(\cdot)$ on $V(G)$ such that $I_{\mathcal{P}}(S)$ consists of the set S itself and every vertex on any path of $\mathcal{P}$ with both ends in S.

Lemma 2.2 *Given a graph G and a family $\mathcal{P}$ of paths in G, we have that $I_{\mathcal{P}}(\cdot)$ is an interval function on $V(G)$, and $C_{\mathcal{P}}$ is a convexity in G.*

Proof Exercise 2.6. □

The most studied graph convexities are path convexities:

- The *geodesic convexity* (Harary and Nieminem 1981)
- The *monophonic convexity* (Jamison 1982)
- the P_3 *convexity* (Centeno et al. 2009)

where $\mathcal{P}$ is the family of every geodesic (shortest path), every induced path, and every P_3 (three-vertex path) of the graph, respectively. Other path convexities in graphs are less known, but have also been the subject of recent research such as:

- The *triangle-path convexity* (Changat and Mathews 1999)
- The m^3 *convexity* (Dragan et al. 1999)
- The P_3^* *convexity* (Araújo et al. 2013)

where $\mathcal{P}$ is the family of paths $v_1 v_2 \ldots$ without edges between vertices v_i and v_j with $|j - i| > 2$, induced paths with at least 3 vertices, and induced P_3's, resp.

Despite the different definitions of these convexities, it may happen that two of them coincide in the same graph. For example, we say that a graph is *distance-hereditary* if every induced path is minimum, like trees. In this graph class, the geodesic and the monophonic convexities coincide, since every minimum path is induced and, in this class, every induced path is minimum. That is, every convex set in the geodesic convexity is also convex in the monophonic convexity, and vice versa, implying that the convexity parameters will also coincide in these graphs.

Another example of *coincidence* is the convexities P_3 and P_3^* in triangle-free graphs, as any P_3 is induced. The geodesic and the P_3^* convexities coincide in diameter 2 graphs, as every geodesic between nonadjacent vertices is an induced P_3. The geodesic and the P_3 convexities also coincide in triangle-free graphs with diameter 2. The lemma below summarizes this information for future use:

Lemma 2.3 *The convexities below are equivalent in the highlighted graphs:*

(a) Geodesic and monophonic in distance-hereditary graphs
(b) Geodesic and P_3^* in graphs with diameter 2
(c) Geodesic and P_3 in triangle-free graphs with diameter 2
(d) P_3 and P_3^* in triangle-free graphs

In this sense, Araújo et al. (2013) show a simple reduction of any graph G to a graph G_u such that the geodesic convexity of G_u is similar to the P_3^* convexity of G. This reduction is shown below and is useful in Chap. 5.

Lemma 2.4 *Given a graph G, let G_u be obtained from G by including a universal vertex u. Then, $S \subseteq V(G)$ is convex in G in the P_3^* convexity if and only if $S \cup \{u\}$ is convex in G_u in the geodesic convexity. In addition, $S \subseteq V(G)$ is convex in G_u in the geodesic convexity if and only if S is a clique of G.*

Proof Exercise 2.6. Note that G_u has diameter 2 and, therefore, every geodesic of G_u between non-neighbors is an induced P_3 of G. □

2.4 Graph Convexities Not Based on Paths

The most popular graph convexities are path convexities. However, other convexities have been studied recently, such as convexities based on threshold functions and convexities based on induced copies of a given graph H. First, we show how to define a graph convexity based on *threshold functions*. As defined by Chen (2009), a *TSS model* (*target set selection*) is given by a graph G and a threshold function $\tau : V(G) \to \mathbb{N}$. The τ-*interval* $I_\tau(S)$ of $S \subseteq V(G)$ is defined as S and every vertex v with at least $\tau(v)$ neighbors in S. Let C_τ be the family of subsets S of $V(G)$ such that $I_\tau(S) = S$, i.e., every vertex v outside S has fewer than $\tau(v)$ neighbors in S. Note that the P_3 convexity can be defined by the threshold function $\tau(v) = 2$ for any vertex v of G. Section 9.1 shows more details about convexities of this type. TSS models are used to represent diffusion processes in graphs, such as information propagation, influence maximization in social networks, and disease contamination.

Lemma 2.5 *Given a graph G with $V(G) \neq \emptyset$ and a threshold function $\tau : V(G) \to \mathbb{N}$, we have that C_τ is a convexity in G if and only if $\tau(v) > 0$ for every $v \in V(G)$.*

Proof Exercise 2.6. Prove when $\mathrm{I}_\tau(\cdot)$ is an interval function. □

Now, we show other way to define convexities in a graph G from induced copies of a given graph H. Let the H-*interval* $\mathrm{I}_H(S)$ of $S \subseteq V(G)$ as S and any vertex v outside S such that $S' \cup \{v\}$ induces a subgraph H in G for some $S' \subseteq S$ (clearly $|S'| = |V(H)|-1$). Let C_H be the family of subsets S of $V(G)$ such that $\mathrm{I}_H(S) = S$, i.e., no vertex outside S induces H with some subset of S.

Lemma 2.6 *Let H be a graph with $|V(H)| \geq 2$. Given a graph G with $V(G) \neq \emptyset$, we have that C_H is a convexity in G.*

Proof Exercise 2.6. Hint: prove that $\mathrm{I}_H(\cdot)$ is an interval function if $|V(H)| \geq 2$. □

Convexity C_H is called *H-free convexity*. Section 4.11 shows that the H-free convexity is a convex geometry in a graph G if and only if G is free of induced H.

Exercises

Exercise 2.1 Prove that the convex hull $\mathrm{conv}_C(\cdot)$ of a convexity C is a *closure operator*.

Exercise 2.2 Provide definitions of convexly independent set, rank, and Carathéodory, Radon, and Helly numbers of a convexity C on a finite set V, as in Chap. 1.

Exercise 2.3 Enumerate all possible convexities on $V = \{a, b\}$ and determine which are convex geometries. Do the same for $V = \{a, b, c\}$.

Exercise 2.4 Given an interval convexity C on V, prove:

(a) $S \subseteq \mathrm{I}_C^k(S) \subseteq \mathrm{conv}_C(S)$, for all $S \subseteq V$ and $k \geq 0$.

(b) $\mathrm{conv}_C(\mathrm{I}_C^k(S)) = \mathrm{conv}_C(S)$, for all $S \subseteq V$ and $k \geq 0$.
(c) $\mathrm{I}_C\left(\bigcup_{k=1}^{\infty} \mathrm{I}_C^k(S)\right) = \bigcup_{k=1}^{\infty} \mathrm{I}_C^k(S)$, for all $S \subseteq V$.
(d) $\mathrm{conv}_C(S) = \bigcup_{k=1}^{\infty} \mathrm{I}_C^k(S)$, for all $S \subseteq V$.

Exercise 2.5 Let $C = \{\emptyset, [n]\}$ be the convexity on $[n] = \{1, \ldots, n\}$ for $n \geq 2$ with interval function $\mathrm{I}_C(\emptyset) = \emptyset$, $\mathrm{I}_C([n]) = [n]$, and $\mathrm{I}_C(S) = S \cup \{\min([n] \setminus S)\}$, if $S \neq \emptyset$ and $S \neq [n]$. (a) Prove that I_C is indeed an interval function. (b) Determine the iteration time and the general position number of $[n]$ in this convexity.

Exercise 2.6 Prove Lemmas 2.2 to 2.6.

Exercise 2.7 State and prove a modification of Lemma 2.6 in the case where H consists of only 1 vertex.

Chapter 3
Graph Convexity Parameters

In this chapter, we focus on the ten most studied graph convexity parameters, listed in Sect. 2.2. There is a subsection to each of them where we recall their definition, list results from the literature, and show examples for simple graphs, determining their values in the most known convexities: geodesic, monophonic, and P_3. Remember that the geodesic, monophonic, and P_3 convexities are associated with minimum paths, induced paths, and P_3 paths within the graph, respectively. As in Lemma 2.3, they can coincide in certain graph classes, such as the geodesic and monophonic convexities in distance–hereditary graphs, in trees, and in the graph of Fig. 3.2.

We assume that every graph G is finite with $V(G) \neq \emptyset$. For graph notations, see Appendix A. Among the simple graphs constantly cited here, the complete graphs K_n, the paths P_n, the cycles C_n, and the trees (acyclic graphs) stand out. We always use n for the number of vertices of the graph. It is known, for example, that $K_3 = C_3$; that every tree has at least 2 leaves, which are vertices of degree 1; and that P_n is a tree with exactly two leaves. In the following, we use the subscripts g, m, and p3 to differentiate the parameters. For example, $\mathrm{h}\ell_{\mathrm{g}}(G)$, $\mathrm{h}\ell_{\mathrm{m}}(G)$, and $\mathrm{h}\ell_{\mathrm{p3}}(G)$ refer to the Helly number of the graph G in the geodesic, monophonic, and P_3 convexities, respectively. When we do not use the subscript, we are referring generically to the parameter, which will depend on the considered convexity.

3.1 Hull Number

The *hull number* $\mathrm{hn}(G)$ is the size of the smallest hull set of G, i.e., $\mathrm{conv}(S) = V(G)$. As stated above, this definition depends on the considered convexity (note that there is no subscript). We will not repeat this in the following sections.

The hull number is one of the most studied convexity parameters and was introduced by Everett and Seidman (1985) in the geodesic convexity. Interestingly,

J. Araújo et al., *Introduction to Graph Convexity*, SBMAC Collection on Applied and Computational Mathematics, https://doi.org/10.1007/978-3-031-84128-6_3

it was not the first to be studied. For example, the iteration time was introduced earlier (Harary and Nieminem 1981), being one of the least investigated. There are several computational complexity results for the hull number in the literature, such as the proof that it is NP-hard in the P_3 convexity even in subgraphs of grids and APX-hard in the geodesic and P_3 convexities (Araújo et al. 2018).

Remember that $\mathrm{Ext}(G)$ is the set of extreme points of $V(G)$. Clearly, every vertex $v \in \mathrm{Ext}(G)$ must be in any hull set, as $V(G) \setminus \{v\}$ is not a hull set since $v \notin \mathrm{conv}(V(G) \setminus \{v\})$. Consequently, $\mathrm{hn}(G) \geq |\,\mathrm{Ext}(G)|$.

The simple lemma below shows that every simplicial vertex (the neighborhood forms a clique) must be in any hull set in two important convexities. In the P_3 convexity on the complete graph K_n for $n \geq 2$, it is easy to see that $\mathrm{hn}_{\mathrm{p3}}(K_n) = 2$.

Lemma 3.1 (Everett and Seidman 1985) *In the geodesic and monophonic convexities, a vertex is an extreme point if and only if it is simplicial (the neighborhood forms a clique). Consequently,* $\mathrm{hn}_{\mathrm{g}}(K_n) = \mathrm{hn}_{\mathrm{m}}(K_n) = n$.

Proof Exercise 3.1. Every vertex is simplicial in the complete graph K_n. □

Applying Lemma 3.1 to the graph G_1 of Fig. 3.1, note that the simplicial vertices v_5, v_6, v_8, and v_9 of G_1 must be in any hull set in the geodesic and monophonic convexities. In the geodesic convexity, $\{v_1, v_5, v_6, v_8, v_9, v_{12}\}$ is a hull set of G_1 of size 6 and there is no smaller one: therefore, $\mathrm{hn}_{\mathrm{g}}(G) = 6$. In the monophonic convexity, $\{v_5, v_6, v_8, v_9, v_{12}\}$ is a hull set of G_1 of size 5 and there is no smaller one: therefore, $\mathrm{hn}_{\mathrm{m}}(G) = 5$. In the P_3 convexity, $\{v_5, v_8, v_{12}, v_{14}\}$ is a hull set of G_1 of size 4 and there is no smaller one: therefore, $\mathrm{hn}_{\mathrm{p3}}(G) = 4$ (Exercise 3.2).

Similarly, the simplicial vertices of a tree are exactly its leaves. Moreover, every non-leaf vertex is on a geodesic between two leaves. Therefore:

Lemma 3.2 *For every tree T,* $\mathrm{hn}_{\mathrm{g}}(T) = \mathrm{hn}_{\mathrm{m}}(T) =$ *number of leaves of T.*

Every path P_n with $n \geq 2$ is a tree with 2 leaves and, therefore, $\mathrm{hn}_{\mathrm{g}}(P_n) = \mathrm{hn}_{\mathrm{m}}(P_n) = 2$. In a cycle C_n with $n \geq 3$ and vertices $v_1 v_2 \ldots v_n$, there are no simplicial vertices, but it is easy to see that any 2 nonadjacent vertices form a hull set in the monophonic convexity. Thus, $\mathrm{hn}_{\mathrm{m}}(C_3) = 3$ and $\mathrm{hn}_{\mathrm{m}}(C_n) = 2$ for $n \geq 4$. Furthermore, in the geodesic convexity, two vertices are sufficient if n is even and

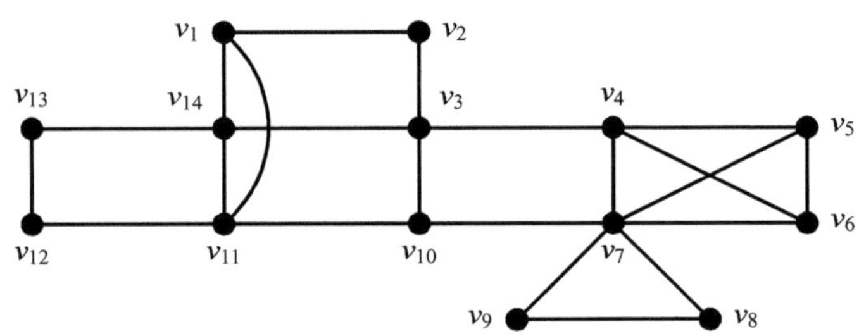

Fig. 3.1 Graph G_1 used in several examples of this chapter

G_2

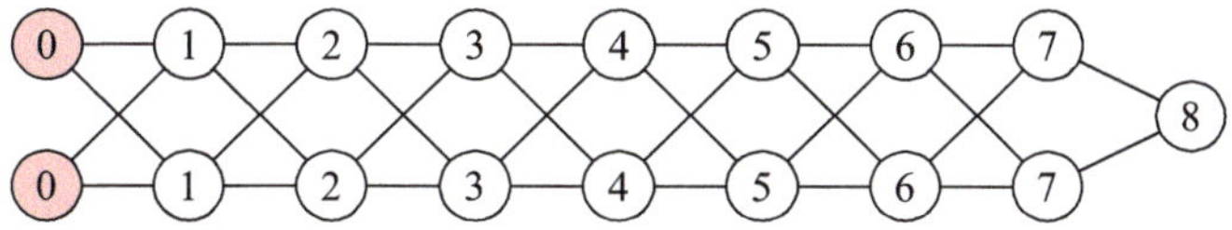

G_3

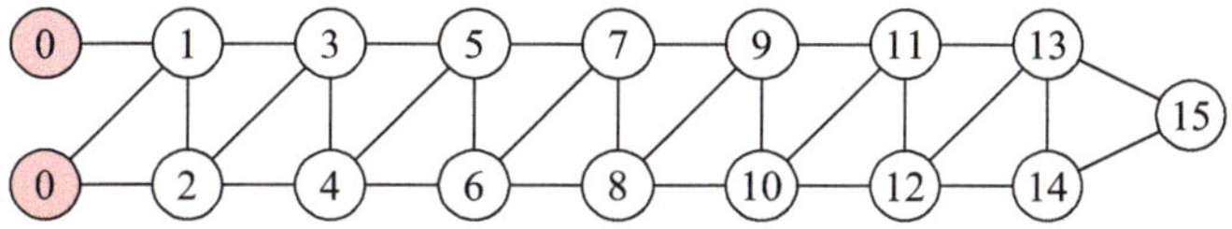

Fig. 3.2 Graphs G_2 and G_3 used as examples in this chapter. The numbers indicate the iteration time in the P_3 convexity of each vertex, starting with the red ones. Note that G_2 is a distance–hereditary graph, but G_3 is not

three vertices are necessary and sufficient if n is odd. Thus, $\mathrm{hn_g}(C_n) = 2$ if n is even and $\mathrm{hn_g}(C_n) = 3$ if n is odd. In the P_3 convexity, we leave as Exercise 3.3 the calculation of $\mathrm{hn_{p3}}(T)$ for any tree T (see Algorithm TSS-SIZE-TREE of Sect. 9.1). The lemma below deals with paths and cycles in the P_3 convexity.

Lemma 3.3 *In the P_3 convexity, every hull set of the path P_n or the cycle C_n, with vertices $v_1, \ldots, v_n$, must contain v_i or v_{i+1} for every $i \in [n]$. Therefore,* $\mathrm{hn_{p3}}(P_n) = \lfloor \frac{n}{2} \rfloor + 1$ *and* $\mathrm{hn_{p3}}(C_n) = \lfloor \frac{n-1}{2} \rfloor + 1$.

Proof Exercise 3.4. □

To conclude, note that the graph G_2 of Fig. 3.2 has no simplicial vertex and that the two red vertices form a hull set in 3 convexities: geodesic, monophonic, and P_3. Therefore, $\mathrm{hn_g}(G_2) = \mathrm{hn_m}(G_2) = \mathrm{hn_{p3}}(G_2) = 2$. Furthermore, in the graph G_3 of Fig. 3.2, the vertices with labels 0 and 15 are simplicial and form a hull set in the geodesic and monophonic convexities. Therefore, $\mathrm{hn_g}(G_3) = \mathrm{hn_m}(G_3) = 3$. In the convexity P_3, the vertices with label 0 form a hull set and, therefore, $\mathrm{hn_{p3}}(G_3) = 2$.

3.2 Interval and Convexity Numbers

Interval Number

The *interval number* $\mathrm{in}(G)$ is the size of the smallest interval set of the graph G. Basically, they are hull sets with iteration time 1. As an interval set is also a hull set, then $\mathrm{hn}(G) \leq \mathrm{in}(G)$. The interval number was introduced by Harary et al. (1993), in the geodesic convexity, under the name of *geodesic number*. In the monophonic convexity, the interval number is also called *monophonic number* (Pelayo 2013).

For the complete graph K_n with at least $n \geq 2$ vertices, it is easy to see that $\text{in}_\text{g}(K_n) = \text{in}_\text{m}(K_n) = n$ and that $\text{in}_\text{p3}(K_n) = 2$.

For the graph G_1 of Fig. 3.1, $\text{in}_\text{g}(G_1) = \text{hn}_\text{g}(G_1) = 6$ and $\text{in}_\text{m}(G_1) = \text{hn}_\text{m}(G_1) = 5$. In the P_3 convexity, it is easy to check that $\{v_2, v_4, v_7, v_8, v_{11}, v_{13}\}$ is an interval set of G_1 of size 6, there being no smaller one and, therefore, $\text{in}_\text{p3}(G_1) = 6 > \text{hn}_\text{p3}(G_1) = 4$.

For the graph G_2 of Fig. 3.2, in the geodesic and monophonic convexities, it is easy to check that the vertices labeled 0 plus the vertex labeled 8 form an interval set of size 3, there being no smaller one and, therefore, $\text{in}_\text{g}(G_2) = \text{in}_\text{m}(G_2) = 3$, which is greater than $\text{hn}_\text{g}(G_2) = \text{hn}_\text{m}(G_2) = 2$. In the P_3 convexity, it is easy to check that the six vertices labeled with 1, 4, and 7 form an interval set of G_2, there being no smaller one and, therefore, $\text{in}_\text{p3}(G_2) = 6$, which is much greater than $\text{hn}_\text{p3}(G_2) = 2$.

The graph G_3 of Fig. 3.2 has 3 simplicial vertices (with labels 0 and 15) that form an interval set in the geodesic and monophonic convexities and, therefore, $\text{in}_\text{g}(G_3) = \text{in}_\text{m}(G_3) = 3$. In the P_3 convexity, it is easy to check that the seven vertices with labels multiple of 3 form an interval set of G_3, there being no smaller one and, therefore, $\text{in}_\text{p3}(G_3) = 7$, which is much greater than $\text{hn}_\text{p3}(G_3) = 2$.

For trees and cycles in the geodesic and monophonic convexities, it is known that hull sets are also interval sets, as their iteration time is 1. Therefore, $\text{in}_\text{g}(T)$ and $\text{in}_\text{m}(T)$ are equal to the number of leaves for every tree T. Furthermore, $\text{in}_\text{g}(C_n) = \text{hn}_\text{g}(C_n)$ and $\text{in}_\text{m}(C_n) = \text{hn}_\text{m}(C_n)$. See Sect. 3.1 for the exact values in cycles.

Convexity Number

The *convexity number* $\text{con}(G)$ is the size of the largest proper convex set of the graph G, i.e., different from $V(G)$. On the other hand, $n - \text{con}(G)$ represents the size of the smallest *coconvex* set (whose complement is convex), basically a set of vertices unreachable by the others. This parameter was introduced by Chartrand et al. (2002b).

There are several computational complexity results for the convexity number. For example, the proof that, besides being NP-hard, it is highly inapproximable in the geodesic and P_3 convexities, i.e., it does not have a polynomial algorithm with approximation factor $n^{1-\varepsilon}$ for any $\varepsilon > 0$, unless P=NP (Coelho et al. 2015).

In the geodesic and monophonic convexities, we have, by Lemma 3.1, that $\text{con}(G) = n - 1$ if and only if G has a simplicial vertex (just take all vertices except a simplicial vertex). Therefore, for the complete graph, $\text{con}_\text{g}(K_n) = \text{con}_\text{m}(K_n) = n - 1$. Furthermore, for the graph G_1 of Fig. 3.1, $\text{con}_\text{g}(G_1) = \text{con}_\text{m}(G_1) = 14 - 1 = 13$, since G_1 has a simplicial vertex. In the P_3 convexity, a vertex is extreme if and only if it has degree 1, i.e., every vertex of degree 1 is in every hull set. So $\text{con}_\text{p3}(G) = n - 1$ if and only if G has a vertex of degree 1 (just take all vertices except a vertex of degree 1). Therefore, $\text{con}_\text{p3}(G_1) \leq 14 - 2 = 12$, since G_1 does not have a vertex of degree 1. As $\{v_{12}, v_{13}\}$ is a coconvex set, then $\text{con}_\text{p3}(G) = 12$.

On the other hand, in the complete graph, $\text{con}_{\text{p3}}(K_n) = 1$. From the above, every tree T with at least 2 vertices has the convexity number equal to $n-1$ in the three convexities, i.e., $\text{con}_{\text{g}}(T) = \text{con}_{\text{m}}(T) = \text{con}_{\text{p3}}(T) = n-1$, since it is enough to take all vertices except a leaf, which is both simplicial and has degree 1.

In a cycle C_n with $n \geq 3$, we have already seen that, in the monophonic convexity, any 2 nonadjacent vertices form an interval set. Therefore, $\text{con}_{\text{m}}(C_n) = 2$ if $n \geq 3$. In the geodesic convexity, it is easy to see that $\text{con}_{\text{g}}(C_n) = \lfloor \frac{n-1}{2} \rfloor$ (all vertices in less than half the cycle). In the P_3 convexity, two adjacent vertices form a coconvex set in C_n for $n \geq 4$ and, therefore, $\text{con}_{\text{p3}}(C_n) = n-2$ if $n \geq 4$ and $\text{con}_{\text{p3}}(C_3) = 1$.

Regarding the graph G_2 of Fig. 3.2, Exercise 3.5 asks for the value of $\text{con}(G_2)$ in the geodesic, monophonic, and P_3 convexities. An important observation is that any two vertices with the same label in G_2 form a hull set and, therefore, no proper convex set can contain two vertices with the same label.

Regarding the graph G_3 of Fig. 3.2, note that the only vertex of degree 1 forms a coconvex set in the 3 convexities and, therefore, $\text{con}_{\text{m}}(G_3) = \text{con}_{\text{g}}(G_3) = \text{con}_{\text{p3}}(G_3) = n - 1 = 16$.

3.3 Iteration and Percolation Times

The two parameters in this section depend on the following important definition: given a graph G and a convexity on G, the *iteration time* of a set $S \subseteq V(G)$, denoted by $\text{ti}(S)$, is the smallest integer k such that $\text{I}^k(S) = \text{conv}(S)$, i.e., k applications of the interval function of the convexity are required to obtain the convex hull of S.

The *iteration time* $\text{ti}(G)$ is the largest value $\text{ti}(S)$ among all sets $S \subseteq V(G)$. This parameter was introduced by Harary and Nieminem (1981) in the first paper on convexity on general graphs, focused on geodesic convexity. In Parvathy and Vijayakumar (1998), bounds were obtained for the geodesic iteration time. Dourado et al. (2016b) obtained an algorithm with complexity $O(n^3 m)$ for distance–hereditary graphs in the geodesic convexity. Moscarini (2020) presented an algorithm of the same complexity for the class containing the graphs in which the geodesic and monophonic convexities coincide. Recently, Araújo et al. (2025) proved that the iteration time is NP-hard in the geodesic and P_3 convexities.

The *percolation time* $\text{tp}(G)$ is the largest value $\text{ti}(S)$ among all sets $S \subseteq V(G)$ that are hull sets of G in the considered convexity. The percolation time was introduced by Benevides and Przykucki (2013) to solve a problem proposed by Béla Bollobás regarding the P_3 convexity. In this convexity, Benevides et al. (2015) proved that the percolation time is NP-hard even in planar graphs and that it is NP-complete to decide if $\text{tp}_{\text{p3}}(G) \geq 4$. In the geodesic convexity, Benevides et al. (2016) obtained a polynomial algorithm in distance–hereditary graphs and proved that it is NP-complete to decide if $\text{tp}_{\text{g}}(G) \geq 2$ even in bipartite graphs.

Consider initially the graphs G_2 and G_3 of Fig. 3.2. It is easy to see directly in the figure that $\text{ti}(G_2) = \text{tp}(G_2) = 8$ in the three convexities: geodesic, monophonic,

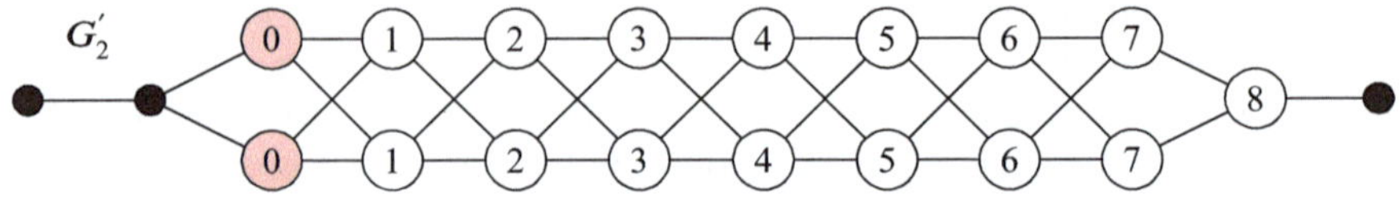

Fig. 3.3 Graph G_2' used as an example to show the difference between iteration time and percolation time. The numbers indicate the iteration time in the P_3 convexity of each vertex, starting with the red ones. Note that the red vertices do not form a hull set

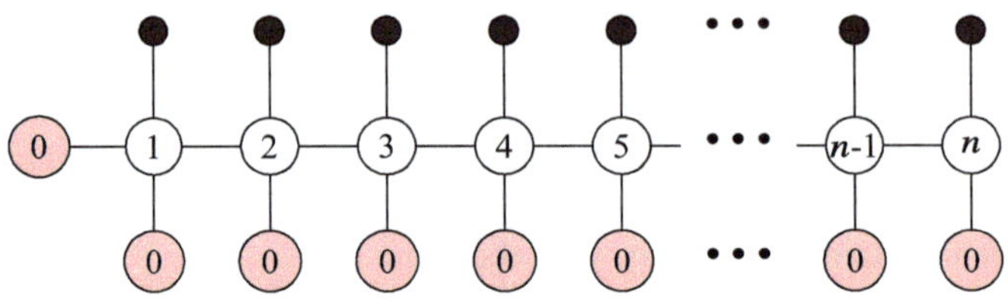

Fig. 3.4 Tree with iteration time n and percolation time 1 in the convexity P_3. The numbers on the vertices represent the iteration time starting at the red vertices

and P_3. In addition, $\text{ti}_{\text{p3}}(G_3) = \text{tp}_{\text{p3}}(G_3) = 15$. The sets with these iteration times are indicated in red in the figure. Note that these sets are also hull sets.

In these examples, the time parameters have the same value, but they can be quite different in other graphs. As an example, consider the graph G_2' of Fig. 3.3 obtained from the graph G_2 of Fig. 3.2 with the inclusion of three new vertices (in black). Observe that the vertices in red do not form a hull set in G_2', but they continue with the same iteration time of 8 and, therefore, $\text{ti}(G_2') = 8$ in the three convexities: geodesic, monophonic, and P_3. Regarding the percolation time, the black vertices of degree 1 must be in the hull set of the three convexities. Since these two vertices of degree 1 also form an interval set in the geodesic and monophonic convexities, then $\text{tp}_{\text{m}}(G_2') = \text{tp}_{\text{g}}(G_2') = 1$. With this, we see that the values of these time parameters can be very different. On the other hand, in the convexity P_3, they will continue to be equal, as the two vertices of degree 1 together with the two in red form a hull set and have the same iteration time. Consequently, $\text{tp}_{\text{p3}}(G_2') = \text{ti}_{\text{p3}}(G_2') = 8$.

Regarding simple graphs, it is not difficult to see that the iteration and percolation times will be equal in the graphs K_n, P_n, and C_n. Let $n \geq 4$. In complete graphs, $\text{ti}_{\text{p3}}(K_n) = \text{tp}_{\text{p3}}(K_n) = 1$ and, in the geodesic and monophonic convexities, $\text{ti}(K_n) = \text{tp}(K_n) = 0$. In paths and cycles, note that the iteration and percolation times are equal to 1 in the three convexities: geodesic, monophonic, and P_3.

It is also not difficult to see that, in trees with more than 2 vertices, the iteration and percolation times are equal to 1 in the geodesic and monophonic convexities. Regarding trees in the convexity P_3, the situation is different and the values of these time parameters can be very different. Figure 3.4 shows, for each $n \geq 1$, a tree with iteration time n and percolation time 1.

Exercise 3.6 asks for the values of the iteration and percolation times in the graphs G_1 and G_3 in the geodesic and monophonic convexities.

3.4 Carathéodory, Radon, and Helly Numbers

The parameters of this section are widely studied in the literature and their definitions were inspired by the classic theorems of Carathéodory, Radon and Helly, presented in Sect. 1.2. Coelho et al. (2015) and Dourado and da Silva (2017) proved that, in addition to being NP-hard, these three parameters are highly inapproximable in the P_3 and geodesic convexities. That is, they do not have a polynomial time algorithm with an approximation factor of $n^{1-\varepsilon}$ for any $\varepsilon > 0$, unless P=NP.

Carathéodory Number

The *Carathéodory number* of a graph G, denoted by $\mathrm{cth}(G)$, is the smallest integer $r \geq 0$ such that, for every subset $S \subseteq V(G)$ and every vertex $u \in \mathrm{conv}(S)$, there exists a subset $F \subseteq S$ with $|F| \leq r$ and $u \in \mathrm{conv}(F)$. This is the closest definition to the classic Theorem 1.3 of Carathéodory (1911).

Alternatively, according to Van de Vel (1993), it is not difficult to prove that the Carathéodory number of G is also the size of the largest *Carathéodory-independent subset*[1] $S \subseteq V(G)$, which are the sets such that $\mathrm{conv}(S) \neq \bigcup_{u \in S} \mathrm{conv}(S \setminus \{u\})$.

Note that as $\mathrm{conv}(S) \supseteq \mathrm{conv}(S \setminus \{u\})$ for every $u \in S$, then $\mathrm{conv}(S) \supseteq \bigcup_{u \in S} \mathrm{conv}(S \setminus \{u\})$. Also note that, in the first definition, the Carathéodory number is a minimization parameter and that, in the second definition, it is a maximization parameter. We leave as Exercise 3.7 the proof that these definitions are also equivalent in abstract convexity.

An example of a maximum Carathéodory-independent set in the P_3 convexity is the set S of the leaves of a complete binary tree B. In each iteration of the interval function $\mathrm{I}_{\mathrm{p3}}(\cdot)$, starting from S, a level of B will be generated in the direction from the leaves to the root, until the root is generated last. That is, $\mathrm{conv}_{\mathrm{p3}}(S)$ contains the root of B, but $\bigcup_{u \in S} \mathrm{conv}_{\mathrm{p3}}(S \setminus \{u\})$ does not contain the root, meaning that all leaves are necessary to generate the root. Thus, $\mathrm{cth}_{\mathrm{p3}}(B) = \mathrm{hn}_{\mathrm{p3}}(B)$ is the number of leaves of B.

In the geodesic convexity, Lemma 2.4 obtains the graph B_u by adding a universal vertex u to the complete binary tree B of the last paragraph. The application of $\mathrm{I}_{\mathrm{g}}(\cdot)$ in B_u, starting from the set S of the leaves of B, will be almost the same as shown for B in the P_3 convexity, with the root being generated last. The difference is that u will be generated in the first iteration, but, as it is universal, u does not help to generate other vertices in the geodesic convexity. Thus, $\mathrm{cth}_{\mathrm{g}}(B_u) = \mathrm{hn}_{\mathrm{g}}(B_u)$ is the number of leaves of B.

[1] Term used by Van de Vel (1993) and Bryant et al. (1978). These sets are also called *irredundant* by Pelayo (2013) and Duchet (1988).

We leave it as Exercise 3.8 to obtain, in the monophonic convexity, a graph with Carathéodory number k for any positive integer k.

Consider the graphs K_n, P_n, and C_n for $n \geq 4$. In complete graphs, $\mathrm{cth_g}(K_n) = \mathrm{cth_m}(K_n) = 1$ and $\mathrm{cth_{p3}}(K_n) = 2$. In cycles, $\mathrm{cth_g}(C_n) = \mathrm{cth_m}(C_n) = \mathrm{cth_{p3}}(C_n) = 2$. In paths, $\mathrm{cth_g}(P_n) = \mathrm{cth_m}(P_n) = \mathrm{cth_{p3}}(P_n) = 2$.

Finally, for the graph G_1 of Fig. 3.1 with $S = \{v_{11}, v_{13}\}$ in the geodesic convexity, note that $\bigcup_{u \in S} conv_g(S\backslash\{u\}) = \{v_{11}, v_{13}\}$ and $conv_g(S) = \{v_{11}, v_{12}, v_{13}, v_{14}\}$. That is, $\{v_{11}, v_{13}\}$ is Carathéodory independent in the geodesic convexity. Furthermore, it is easy to see that every set with 3 or more vertices is Carathéodory dependent and, therefore, $\mathrm{cth_g}(G_1) = 2$ (Exercise 3.9). Moreover, every Carathéodory-dependent set of G_1 in the geodesic convexity will also be in the monophonic convexity and, therefore, $\mathrm{cth_m}(G_1) = 2$. In the P_3 convexity on G_1, we have $\mathrm{cth_{p3}}(G_1) = 4$. As examples of maximum Carathéodory-independent sets in the P_3 convexity, we have $\{v_1, v_2, v_5, v_{11}\}$ and $\{v_6, v_9, v_{12}, v_{14}\}$. We leave as Exercise 3.9 to show that every set with 5 or more vertices of G_1 is Carathéodory dependent in the P_3 convexity.

We leave as Exercise 3.10 to determine $\mathrm{cth}(G_2)$ and $\mathrm{cth}(G_3)$ in the geodesic, monophonic, and P_3 convexities.

Radon Number

A set $S \subseteq V(G)$ is *Radon dependent* if there exists a partition of S into two sets S_1 and S_2 satisfying $\mathrm{conv}(S_1) \cap \mathrm{conv}(S_2) \neq \emptyset$; otherwise, it is *Radon independent*. The *Radon number* $\mathrm{rd}(G)$ is the size of the largest Radon-independent set[2] of G. It is easy to see that the Radon-independent property is hereditary, i.e., if $A \subseteq B$ and B is Radon independent, then A is also Radon independent.

Note initially that every clique is a Radon-independent set in the geodesic and monophonic convexities. Therefore, in the complete graph, $\mathrm{rd_g}(K_n) = \mathrm{rd_m}(K_n) = n$. Furthermore, it is easy to see that $\mathrm{rd_{p3}}(K_n) = 2$ for all $n \geq 3$. In addition, in paths with $n \geq 3$ vertices, $\mathrm{rd_m}(P_n) = \mathrm{rd_g}(P_n) = 2$ and $\mathrm{rd_{p3}}(P_n) = \left\lceil \frac{2n}{3} \right\rceil$, because the set cannot have a P_3. Finally, in cycles with $n \geq 4$ vertices, $\mathrm{rd_m}(C_n) = 2$, $\mathrm{rd_g}(C_n) = 3$ if $n \geq 5$, $\mathrm{rd_g}(C_4) = 2$ and $\mathrm{rd_{p3}}(C_n) = \left\lfloor \frac{2n}{3} \right\rfloor$.

Now consider the graph G_1 of Fig. 3.1. Since the maximum clique of G_1 has 4 vertices, then $\mathrm{rd_g}(G_1) \geq 4$. We leave as Exercise 3.11 the proof that every set with 5 vertices is Radon dependent in the geodesic convexity. Therefore, $\mathrm{rd_g}(G_1) = 4$. Note that every Radon-dependent set in the geodesic convexity is also Radon dependent in the monophonic convexity. Thus, $\mathrm{rd_m}(G_1) = 4$. Note that $\{v_2, v_8, v_9, v_{12}, v_{14}\}$ is Radon independent in the P_3 convexity, so $\mathrm{rd_{p3}}(G_1) \geq 5$.

[2] According to Van de Vel (1993), some authors define the Radon number as 1 plus the size of the largest Radon-independent set. We follow the definition of Van de Vel (1993).

We leave as Exercise 3.11 the proof that every set with 6 vertices is Radon dependent in the P_3 convexity. Therefore, $\mathrm{rd}_{\mathrm{p3}}(G_1) = 5$.

We leave as Exercise 3.12 the calculation of $\mathrm{rd}(G_2)$ and $\mathrm{rd}(G_3)$ in the geodesic, monophonic, and P_3 convexities.

Finally, consider a completely binary tree B. Note that the set S of leaves of B is Radon independent, since any partition would generate different subtrees. As there is no larger Radon-independent set in B (Exercise 3.13), then $\mathrm{rd}(B)$ is equal to the number of leaves.

Helly Number

According to Van de Vel (1993), the *Helly number* $\mathrm{h}\ell(G)$ is the size of the largest *Helly-independent* subset $S \subseteq V(G)$, which are the sets such that $\bigcap_{v \in S} \mathrm{conv}(S \setminus \{v\}) = \emptyset$.

Note that the Helly-independent property is hereditary, i.e., if $A \subseteq B$ and B is Helly independent, then A is also Helly independent. Also note that $\emptyset$ and $\{v\}$ are Helly independent for every $v \in V(G)$. That is, if $|V(G)| \leq 1$, then $\mathrm{h}\ell(G) = |V(G)|$. Theorem 3.2 shows that $\mathrm{h}\ell(G) \geq 2$ for every graph G with at least two vertices in the main graph convexities.[3]

Alternatively, according to Van de Vel (1993), it was independently proved by Calder (1971), Berge and Duchet (1975), and Sierksma (1975) that, if $\mathrm{h}\ell(G) \geq 2$, the Helly number $\mathrm{h}\ell(G)$ is also the smallest integer $h \geq 2$ such that every family $\mathcal{F}$ of convex subsets of $V(G)$ in which each h members of $\mathcal{F}$ have a common point satisfies the property that all members of $\mathcal{F}$ have a common point. This is the definition closest to the classic Theorem 1.5 of Helly (1923).

Theorem 3.1 (Jamison and Nowakowski 1984, Duchet 1988) *In the monophonic convexity, the Helly number is equal to the size of the largest clique, i.e.,* $\mathrm{h}\ell_{\mathrm{m}}(G) = \omega(G)$ *for every graph G.*

Therefore, in the geodesic convexity, $\mathrm{h}\ell_{\mathrm{g}}(G) = \omega(G)$ for every distance–hereditary graph G (recall Lemma 2.3). For example, in every tree T with at least 2 vertices, $\mathrm{h}\ell_{\mathrm{g}}(T) = \mathrm{h}\ell_{\mathrm{m}}(T) = 2$. Furthermore, in complete graphs, $\mathrm{h}\ell_{\mathrm{g}}(K_n) = \mathrm{h}\ell_{\mathrm{m}}(K_n) = n$ (also note that every set of vertices of K_n is convex and, therefore, $V(K_n)$ is maximum Helly independent). Finally, regarding the graph G_2 of Fig. 3.2, $\mathrm{h}\ell_{\mathrm{g}}(G_2) = \mathrm{h}\ell_{\mathrm{m}}(G_2) = 2$ as it is distance–hereditary.

In the P_3 convexity, $\mathrm{h}\ell_{\mathrm{p3}}(K_n) = 2$ for $n \geq 2$. In cycles, note that $\mathrm{h}\ell_{\mathrm{m}}(C_n) = 2$ for $n \geq 4$, and that $\mathrm{h}\ell_{\mathrm{g}}(C_n) = 3$ for $n \geq 5$ and $\mathrm{h}\ell_{\mathrm{g}}(C_4) = 2$ (Carvalho 2016). We leave as Exercise 3.14 to determine $\mathrm{h}\ell_{\mathrm{p3}}(P_n)$ and $\mathrm{h}\ell_{\mathrm{p3}}(C_n)$.

[3] Convexities in which $\mathrm{h}\ell(G) < 2$ are uncommon and of little interest, like the convexities $\{\emptyset, V(G)\}$ and $\{\emptyset, \{v_1\}, \{v_1, v_2\}, \{v_1, v_2, v_3\}, \ldots\}$

Consider the graph G_1 of Fig. 3.1. From Theorem 3.1, $h\ell_m(G_1) = 4$. Note that every clique is a Helly-independent set in the geodesic convexity. Since the maximum clique of H has 4 vertices, then $h\ell_g(G_1) \geq 4$. We leave it as Exercise 3.15 to show that every set with 5 vertices is Helly dependent in the geodesic convexity in G_1. Therefore, $h\ell_g(G_1) = 4$. In the P_3 convexity, note that $\{v_2, v_8, v_{12}, v_{14}\}$ is Helly independent. Therefore, $h\ell_{p3}(G_1) \geq 4$. We leave as Exercise 3.15 to show that every set with 5 vertices is Helly dependent in the P_3 convexity. Therefore, $h\ell_{p3}(G_1) = 4$. We also leave as Exercise 3.16 to determine $h\ell(G_2)$ and $h\ell(G_3)$ in the P_3 convexity.

3.5 General Position Number and Rank

General Position Number

A subset S of vertices of a graph G is in *general position* in a convexity C if S does not contain 3 elements x, y, z such that z is in the interval of x and y, i.e., $z \notin I_C(\{x, y\})$. The *general position number* $gp_C(S)$ of G is the largest subset of $V(G)$ in general position in C.

The classic problem *No-Three-in-Line* by Dudeney (1917)[4] seeks the largest number of points in the $m \times m$ grid without three collinear points, a problem which remains open for $m > 46$ (Flammenkamp 1998). Motivated by this, Manuel and Klavžar (2018) introduced the general position number $gp_g(G)$ in the geodesic convexity and proved that it is NP-hard. Recently Araújo et al. (2025) proved that the general position number is also NP-hard in the monophonic convexity even in graphs with diameter two.

Interestingly, in the P_3 convexity, any set in general position induces a subgraph with maximum degree 1 and, therefore, $gp_{p3}(G)$ is equivalent to the *dissociation number* diss(G) of the graph G, a parameter introduced by Yannakakis (1981) and proved NP-hard even in bipartite graphs and in planar graphs with maximum degree 4. In the convexity P_3^*, any set in general position induces a subgraph whose connected components are cliques and, therefore, $gp_{p3*}(G)$ is equivalent to the *IUC number* of the graph G (*IUC—independent union of cliques*), a parameter introduced by Ertem et al. (2020) and proved NP-hard even in planar graphs.

Let $n \geq 4$. In complete graphs, $gp(K_n) = n$ in the geodesic and monophonic convexities, as every subset is convex. In the P_3 convexity, $gp_{p3}(K_n) = 2$, as every vertex is in the interval between any other two. In paths and cycles, $gp_m(C_n) = gp_m(P_n) = gp_g(P_n) = 2$, because, for any 3 vertices, one is in the interval of the other two. Regarding cycles in the geodesic convexity, $gp_g(C_n) = 3$ for $n \geq 5$ and $gp_g(C_4) = 2$. In the P_3 convexity, $gp_{p3}(P_n) = \lceil \frac{2n}{3} \rceil$ and

[4] Initially defined as *Puzzle with Pawns* in the book by Dudeney (1917), which was presented by the prestigious journal Nature in the paper Amusements in Mathematics (1917).

$\text{gp}_{\text{p3}}(C_n) = \lfloor \frac{2n}{3} \rfloor$, because the set cannot contain three consecutive vertices. Note that in these examples the general position number obtained the same value as the Radon number. This will be useful in the next section, about the rank of the graph.

For the graph G_1 of Fig. 3.1, note that the set $\{v_1, v_5, v_6, v_8, v_9, v_{12}, v_{13}\}$ is in general position in the geodesic convexity. Therefore, $\text{gp}_{\text{g}}(G_1) \geq 7$. We leave as Exercise 3.17 to show that any set with 8 vertices is not in general position in the geodesic convexity. Therefore, $\text{gp}_{\text{g}}(G_1) = 7$. Note that the set $\{v_5, v_6, v_8, v_9, v_{12}\}$ is in general position in the monophonic convexity; therefore, $\text{gp}_{\text{m}}(G_1) \geq 5$. We leave as Exercise 3.17 to show that any set with 6 vertices of G_1 is not in general position in the monophonic convexity. Therefore, $\text{gp}_{\text{m}}(G_1) = 5$. Note that the set $\{v_1, v_2, v_5, v_6, v_8, v_9, v_{10}, v_{12}, v_{13}\}$ is in general position in the P_3 convexity; therefore, $\text{gp}_{\text{p3}}(G_1) \geq 9$. We leave as Exercise 3.17 to show that any subset of $V(G_1)$ with 10 vertices is not in general position in the P_3 convexity. Therefore, $\text{gp}_{\text{p3}}(G_1) = 9$.

We leave as Exercise 3.18 to determine $\text{gp}(G_2)$ and $\text{gp}(G_3)$ in the P_3, geodesic, and monophonic convexities.

The following lemma is used in Sect. 9.2 and shows a characterization of sets in general position in the geodesic convexity of bipartite graphs. It is in the proof of Theorem 5.1 of Anand et al. (2019).

Lemma 3.4 (Anand et al. 2019) *Let G be a connected bipartite graph with at least 3 vertices. If $S \subseteq V(G)$ is a set in geodesic general position with $|S| \geq 3$, then S is an independent set. Therefore,* $\text{gp}_{\text{g}}(G) \leq \alpha(G)$.

Rank

A subset S of vertices of a graph G is in *convex position* or is *convexly independent* if no $x \in S$ belongs to $\text{conv}(S \setminus \{x\})$; otherwise, it is *convexly dependent*. The *rank* of G, denoted by $\text{rk}(G)$, is defined as the size of the largest convexly independent subset of $V(G)$.

The rank was introduced by Jamison (1981) in the early days of Graph Convexity. In the monophonic and P_3 convexities, Ramos et al. (2014) proved that the rank is NP-hard. In the geodesic convexity, Kanté et al. (2017) proved that the rank is NP-hard even in bipartite graphs and obtained a polynomial algorithm for distance–hereditary graphs.

Section 3.6 shows that $\text{gp}(G) \geq \text{rk}(G) \geq \text{rd}(G) \geq \text{h}\ell(G)$ in the main graph convexities. The following results follow directly from this inequality and the fact that the general position number is equal to the Radon number in the graphs K_n, P_n, and C_n, as seen in the previous section.

Let $n \geq 4$. In the monophonic convexity, $\text{rk}_{\text{m}}(K_n) = n$ and $\text{rk}_{\text{m}}(P_n) = \text{rk}_{\text{m}}(C_n) = 2$. In the geodesic convexity, $\text{rk}_{\text{g}}(K_n) = n$, $\text{rk}_{\text{g}}(P_n) = 2$, $\text{rk}_{\text{g}}(C_n) = 3$

for $n \geq 5$ and $\text{rk}_g(C_4) = 2$. In the P_3 convexity, $\text{rk}_{p3}(K_n) = 2$, $\text{rk}_{p3}(P_n) = \lceil \frac{2n}{3} \rceil$ and $\text{rk}_{p3}(C_n) = \lfloor \frac{2n}{3} \rfloor$.

Consider the graph G_1 of Fig. 3.1. Note that $\{v_1, v_5, v_6, v_8, v_9, v_{12}, v_{13}\}$ is convexly independent in the geodesic convexity, so $\text{rk}_g(G_1) \geq 7$. We leave as Exercise 3.19 the proof that every subset of $V(G_1)$ with 8 vertices is convexly dependent in the geodesic convexity. Therefore, $\text{rk}_g(G_1) = 7$. Note that $\{v_1, v_5, v_6, v_8, v_9\}$ is convexly independent in the monophonic convexity, so $\text{rk}_g(G_1) \geq 5$. We leave as Exercise 3.19 the proof that every subset of $V(G_1)$ with 6 vertices is convexly dependent in the monophonic convexity. Therefore, $\text{rk}_m(G_1) = 5$. Since $\text{rk}_{p3}(G_1) \geq \text{h}\ell_{p3}(G_1) = 4$, then $\text{rk}_{p3}(G_1) \geq 4$. We leave as Exercise 3.19 the proof that every set with 5 vertices is convexly dependent in the P_3 convexity. Therefore, $\text{rk}_{p3}(G_1) = 4$. We leave as Exercise 3.20 the calculation of $\text{rk}(G_2)$ and $\text{rk}(G_3)$ in the geodesic, monophonic, and P_3 convexities.

3.6 Inequalities Between the Parameters

We start with the most basic inequalities, left as an exercise.

Lemma 3.5 *For every graph G and every convexity on G,*

- $\text{in}(G) \geq \text{hn}(G) \geq |\text{Ext}(G)|$
- $\text{ti}(G) \geq \text{tp}(G)$
- $\text{gp}(G) \geq \text{rk}(G)$

Proof Exercise 3.21. □

The inequality below assumes a very common restriction in all major graph convexities: each vertex separately forms a convex set. This holds, for example, for all path convexities, such as P_3, P_3^*, geodesic, monophonic, and triangular.

Theorem 3.2 *For every graph G with at least two vertices and every convexity on G such that $\{v\}$ is convex for every $v \in V(G)$:*

- $\text{h}\ell(G) \geq 2$
- $\text{rk}(G) \geq \text{rd}(G)$

Proof For the first, it is easy to check that any set with two vertices of G is Helly independent. For the second, it is sufficient to prove that every Radon-independent set is convexly independent, because, as they are maximization parameters, the size of the largest convexly independent set will be greater than or equal to the largest Radon-independent set.

Let $S \subseteq V(G)$ be a non-empty Radon-independent set and let $v \in S$. By the statement, $\text{conv}(\{v\}) = \{v\}$. Since S is Radon independent, then $S \setminus \{v\}$ and $\{v\}$ cannot form a partition such that $\text{conv}(S \setminus \{v\}) \cap \text{conv}(\{v\}) \neq \emptyset$. That is, $\text{conv}(S \setminus \{v\}) \cap \{v\} = \emptyset$ and, therefore, $v \notin \text{conv}(S \setminus \{v\})$. As this holds for every $v \in S$, then S is convexly independent. □

The following inequalities hold for any graph convexity. Note that the next inequality is *tight* in complete graphs in the geodesic, monophonic, and P_3 convexities, because $\mathrm{rd_g}(K_n) = \mathrm{h\ell_g}(K_n) = n$ and $\mathrm{rd_{p3}}(K_n) = \mathrm{h\ell_{p3}}(K_n) = 2$.

Theorem 3.3 (Levi 1951) *For every graph G and every convexity on G,*

$$\mathrm{rd}(G) \;\geq\; \mathrm{h\ell}(G)$$

Proof It is sufficient to prove, similar to the previous theorem, that every Helly-independent set is Radon independent, as they are maximization parameters. We will prove the contrapositive of this, i.e., every Radon-dependent set is Helly dependent.

Let $S \subseteq V(G)$ be a Radon-dependent set, i.e., there is a bipartition of S into subsets S_1 and S_2 satisfying $\mathrm{conv}(S_1) \cap \mathrm{conv}(S_2) \neq \emptyset$. Let $p \in \mathrm{conv}(S_1) \cap \mathrm{conv}(S_2)$. For every $v \in S$, note that $S_1 \subseteq S \setminus \{v\}$ or $S_2 \subseteq S \setminus \{v\}$. Therefore, $p \in \mathrm{conv}(S \setminus \{v\})$ for every $v \in S$, which implies that S is Helly dependent, because $\bigcap_{v \in S} \mathrm{conv}(S \setminus \{v\}) \neq \emptyset$. □

The next inequality was obtained by Eckhoff and Jamison and, according to Van de Vel (1993), was communicated in the doctoral thesis of Sierksma (1976). Interestingly, Jamison (1974) also did his doctoral thesis on convexity.

Theorem 3.4 (Eckhoff–Jamison 1976) *For every graph G and every convexity on G, if* $\mathrm{h\ell}(G) \geq 2$, *then*

$$\mathrm{cth}(G) \;\geq\; \frac{\mathrm{rd}(G) - 1}{\mathrm{h\ell}(G) - 1}$$

Proof Let $c = \mathrm{cth}(G)$ and $h = \mathrm{h\ell}(G) \geq 2$. Let $R \subseteq V(G)$ with $|R| \geq c \cdot (h-1) + 2$ and let $p \in R$. Consider the family $\mathcal{F}$ with the set $\mathrm{conv}(R \setminus \{p\})$ and every set $\mathrm{conv}(R \setminus A)$ for $A \subseteq R$ with $|A| \leq c$ such that $p \notin A$. We will prove that each h members of $\mathcal{F}$ have a common point.

Take h members of $\mathcal{F}$. If $\mathrm{conv}(R \setminus \{p\})$ was not selected, then all h members contain the vertex p. On the other hand, if $\mathrm{conv}(R \setminus \{p\})$ was selected, then the remaining $h-1$ members are of the type $\mathrm{conv}(R \setminus A_i)$ for $1 \leq i < h$. Therefore,

$$\left|\bigcup_{i=1}^{h-1} A_i\right| \leq \sum_{i=1}^{h-1} |A_i| \leq c \cdot (h-1) \leq |R| - 2$$

and, therefore, there are at least two vertices of R not covered by the sets A_i. Let $q \neq p$ be one of them. Then q belongs to all h members of $\mathcal{F}$ as desired.

Therefore, by the definition of the Helly number, there is a vertex x common to every member of the family $\mathcal{F}$. Furthermore, by the definition of the Carathéodory number, as $x \in \mathrm{conv}(R \setminus \{p\})$, then there exists $A \subseteq R \setminus \{p\}$ with $|A| \leq c$ such that $x \in \mathrm{conv}(A)$. This means that R is Radon dependent, since it can be partitioned into two sets A and $R \setminus A$ such that $\text{conv}(A) \cap \text{conv}(R \setminus A)$ is not empty as it contains x. Therefore, $\text{rd}(G) \leq c \cdot (h-1) + 1$. □

Note that this inequality is *tight* in complete graphs in the geodesic and monophonic convexities, as $\text{cth}_g(K_n) = 1$ and $\text{rd}_g(K_n) = \text{h}\ell_g(K_n) = n$. An almost tight example occurs in the P_3 convexity in any completely binary tree B, as $\text{h}\ell(B) = 2$ and $\text{cth}(B)$ and $\text{rd}(B)$ are equal to the number of leaves. It was also proved that this inequality is *tight* for other particular cases in Sierksma et al. (2000). See Exercise 3.22. According to Van de Vel (1993), the problem of showing that the Eckhoff–Jamison inequality is tight for general cases was popular in the late 1970s, but was abandoned without a satisfactory answer, despite many efforts. Other inequalities can be seen in Sierksma (1977).

Exercises

Exercise 3.1 Prove Lemma 3.1: a vertex is an extreme point in the geodesic and monophonic convexities if and only if it is simplicial.

Exercise 3.2 Prove that G_1 does not have a hull set of size 3 in the P_3 convexity.

Exercise 3.3 Obtain a polynomial time algorithm to determine $\text{hn}_{p3}(T)$ for any tree T in the P_3 convexity. (Hint: Algorithm TSS-SIZE-TREE of Sect. 9.1).

Exercise 3.4 Prove Lemma 3.3.

Exercise 3.5 Determine $\text{con}(G_2)$ in the geodesic, monophonic, and P_3 convexities.

Exercise 3.6 Determine the values of ti_g, ti_m, tp_g, and tp_m for the graphs G_1 and G_3.

Exercise 3.7 Prove that the definitions of Carathéodory number are equivalent.

Exercise 3.8 Obtain a graph G_k with $\text{cth}_m(G_k) = k$ for any positive integer k.

Exercise 3.9 Show that $\text{cth}_g(G_1) = 2$ and $\text{cth}_{p3}(G_1) = 4$.

Exercise 3.10 Determine the values of cth_g, cth_m, and cth_{p3} for graphs G_2 and G_3.

Exercise 3.11 Show that $\text{rd}_g(G_1) = 4$ and $\text{rd}_{p3}(G_1) = 5$.

Exercise 3.12 Determine the values of rd_g, rd_m, and rd_{p3} for graphs G_2 and G_3.

Exercise 3.13 Let B be a completely binary tree. Prove that there is no Radon-independent set of B with size larger than the number of leaves.

Exercise 3.14 Determine the values of $\text{h}\ell_{p3}(P_n)$ and $\text{h}\ell_{p3}(C_n)$.

Exercise 3.15 Show that $\text{h}\ell_g(G_1) = 4$ and $\text{h}\ell_{p3}(G_1) = 4$.

Exercise 3.16 Determine the values of $\text{h}\ell_{p3}(G_2)$ and $\text{h}\ell_{p3}(G_3)$ in the P_3 convexity.

Exercise 3.17 Show that $\mathrm{gp_g}(G_1) = 7$, $\mathrm{gp_m}(G_1) = 5$, and $\mathrm{gp_{p3}}(G_1) = 9$.

Exercise 3.18 Determine the values of $\mathrm{gp_g}$, $\mathrm{gp_m}$, and gp_{pc} for graphs G_2 and G_3.

Exercise 3.19 Show that $\mathrm{rk_g}(G_1) = 7$, $\mathrm{rk_m}(G_1) = 5$, and $\mathrm{rk_{p3}}(G_1) = 4$.

Exercise 3.20 Determine the values of $\mathrm{rk_g}$, $\mathrm{rk_m}$, and $\mathrm{rk_{p3}}$ for graphs G_2 and G_3.

Exercise 3.21 Prove the basic inequalities of Lemma 3.5.

Exercise 3.22 Find at least one graph in which the Eckhoff–Jamison inequality of Theorem 3.4 is tight, for each one of the main graph convexities: geodesic, monophonic, and P_3.

Chapter 4
Convex Geometries in Graphs

As seen in Sect. 2.1, we say that a convexity C on a finite set V is a *convex geometry* (or a *geometric convexity*) if it satisfies the *Minkowski–Krein–Milman property*: every convex set is the convex hull of its extreme points. One of the main tools used to prove that a certain convexity C is a convex geometry is to verify the *anti-exchange property* or *antimatroid property*: for every convex set $S \subseteq V$ and $z_1, z_2 \notin S$, it holds that $z_1 \in \text{conv}_C(S \cup \{v_2\})$ implies $z_2 \notin \text{conv}_C(S \cup \{v_1\})$.

Lemma 4.1 (Edelman and Jamison 1985) *Let C be a convexity on a finite set V. Then C is a convex geometry if and only if it satisfies the anti-exchange property.*

Proof Exercise 4.1. □

As defined above, the anti-exchange property is a combinatorial abstraction of the convex hull in a space. Figure 4.1 shows an example in the space $\mathbb{R}^2$. The points z_1 and z_2 do not belong to the convex hull of S. The point z_1 is in the convex hull of $S \cup \{z_2\}$, but the point z_2 does not belong to the convex hull of $S \cup \{z_1\}$. We use here an equivalent formulation of the anti-exchange property, in which S is not necessarily convex: if $z_1, z_2 \notin \text{conv}(S)$ and $z_1 \in \text{conv}(S \cup \{z_2\})$, then $z_2 \notin \text{conv}(S \cup \{z_1\}$.

As we will see, we can prove that a certain graph convexity is a convex geometry by directly applying the Minkowski–Krein–Milman property or using, as an alternative tool, the anti-exchange property.

J. Araújo et al., *Introduction to Graph Convexity*, SBMAC Collection on Applied and Computational Mathematics, https://doi.org/10.1007/978-3-031-84128-6_4

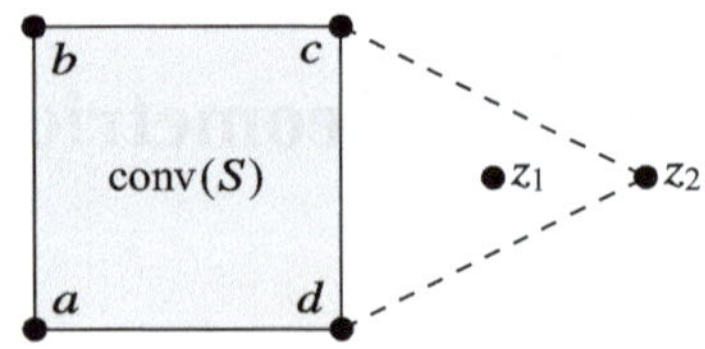

Fig. 4.1 Anti-exchange property applied to $S = \{a, b, c, d\}$

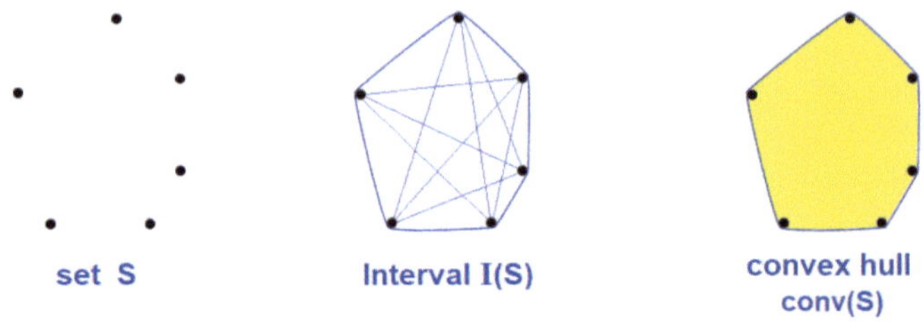

Fig. 4.2 Repeating Fig. 1.2

4.1 Path-Based Convex Geometries

In Chap. 1, we defined the interval function $\mathrm{I}(\cdot)$ for sets in the d-dimensional space. We observed that this function can be used to determine the convex hull of a set S through successive applications of it. See Fig. 4.2, which reproduces Fig. 1.2 again. Given a set S of points in space (represented on the left in the figure), the first application of the interval function results in the set $S_1 = \mathrm{I}(S)$ (represented in the center). We can now imagine reapplying the interval function to include new points that belong to segments whose endpoints are in S_1, obtaining a new set $S_2 = \mathrm{I}(S_1)$ (represented on the right in the figure). Note that $S \subseteq S_1 \subseteq S_2$. Since S_2 is convex, the process ends, as $S_3 = \mathrm{I}(S_2) = S_2$. We can then naturally imagine an iterative process in which we apply the function $\mathrm{I}(\cdot)$ several times until we obtain a convex set S_k. This set satisfies the Minkowski–Krein–Milman property, i.e., S_k is the convex hull of its extreme points (in Fig. 4.2, such extreme points are the points of S itself, but this is not always the case).

Moving from the context of geometry to that of combinatorics, we mentioned in Chap. 2 that the interval function $\mathrm{I}_C(\cdot)$, for a convexity C in a graph G, is the combinatorial analogue of the interval function $\mathrm{I}(\cdot)$. Let us use an example to better understand this statement.

Suppose that $\mathcal{P}$ is the family of all shortest paths of a graph G and C is the family formed by every subset $S \subseteq V(G)$ such that S contains every vertex of any shortest path with endpoints in S. By Lemma 2.2, C is a path-based convexity in G. Specifically, C is the geodesic convexity of G, introduced in Sect. 2.3. In this case, we define $\mathrm{I}_C(S)$, for $S \subseteq V(G)$, as the set containing every vertex that is in S or on a shortest path between two vertices of S. Note that this definition of $\mathrm{I}_C(S)$ follows what we would naturally understand as a combinatorial abstraction of *taking all points that are on segments with endpoints in S*.

Fig. 4.3 Gem graph

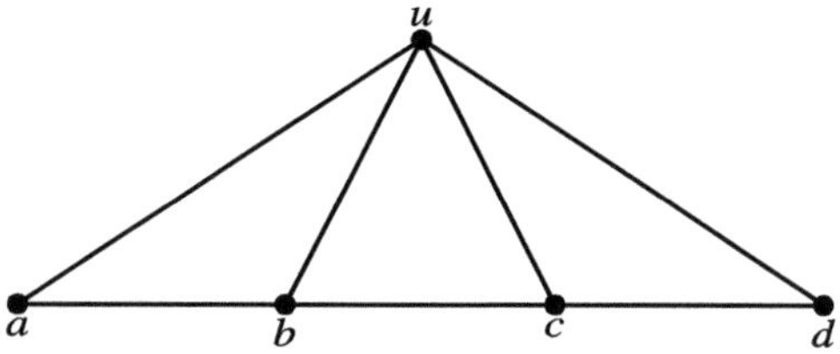

For $S \subseteq V(G)$, let $S = S_0$ and $S_{k+1} = \mathrm{I}_C(S_k)$, $k \geq 0$. Note that this defines an iterative process of generating sets satisfying $S_0 \subseteq S_1 \subseteq S_2 \ldots$. Since G is a finite graph, we will reach a situation where it is no longer possible to include new vertices in S_k, i.e., $S_{k+1} = S_k$. Therefore, S_k is convex. Moreover, it is easy to see that $S_k = \mathrm{conv}_C(S)$ (Exercise 4.2). This process is exactly the combinatorial parallel of the geometric process described at the beginning of this section.

The natural question that arises, then, is: Is it true that the convex set $S_k = \mathrm{conv}_C(S)$ obtained by the iterative application of the function $\mathrm{I}_C(\cdot)$ also satisfies the Minkowski–Krein–Milman property, as in geometry? The answer is no. A simple example elucidates this issue. Remember that $\mathrm{Ext}_C(G) = \mathrm{Ext}_C(V(G))$ and that the subscript g refers to the geodesic convexity.

Example 4.1 The *gem* is the graph G represented in Fig. 4.3 with $V(G) = \{a, b, c, d, u\}$. In the geodesic convexity, the extreme points (vertices) of G are a and d. This fact can be seen as follows: vertex a is not an internal vertex of any shortest path with endpoints at two other vertices of $V(G)$, as its neighbors (b and u) are adjacent. Therefore, $a \notin \mathrm{conv}_{\mathrm{g}}(V(G) \setminus \{a\})$. The same occurs for vertex d. In other words, $\mathrm{Ext}_{\mathrm{g}}(G) = \{a, d\}$. On the other hand, $\mathrm{conv}_{\mathrm{g}}(\{a, d\}) = \{a, d, u\}$. To show this fact, apply successive applications of the interval function. Making $S_0 = S = \{a, d\}$, note that $S_1 = \mathrm{I}_{\mathrm{g}}(S_0) = \{a, d, u\}$, because $P = aud$ is the only shortest path with endpoints a and d. Applying the interval function again, it is easy to see that $S_2 = \mathrm{I}_{\mathrm{g}}(S_1) = \{a, d, u\} = S_1$. Thus, indeed, $\mathrm{conv}_{\mathrm{g}}(\{a, d\}) = \{a, d, u\}$. But, as $S_2 \neq V(G)$, note that $V(G)$ is *not* the convex hull of its extreme vertices, i.e., the Minkowski–Krein–Milman property fails in this case.

The previous example leads to the following question:

Given a rule for determining the convex sets of a convexity C, what are the graphs G for which the convexity C of G is a convex geometry?

For example, the monophonic convexity of a graph G is a convex geometry if and only if G is a chordal graph; and the geodesic convexity of G is a convex geometry if and only if G is Ptolemaic.

Table 4.1 lists some of the most important results of this type. In the next sections, we will describe these results in detail.

Table 4.1 Overview of results of the type: *a graph G belongs to the class $\mathcal{G}$ if and only if the convexity C of G is a convex geometry*

Graph class $\mathcal{G}$	Associated convex geometry C
Chordal	Monophonic
Ptolemaic	Geodesic
Acyclic	Triangle path
Star forests	P_3
Chordal cographs	P_3^* (or l^2)
Strongly chordal	Strong
Weakly polarizable	m^3
Interval graphs	Toll convexity
Unit interval graphs	Weakly toll convexity

4.2 Monophonic Convexity and Chordal Graphs

A graph G is *chordal* if G does not contain any cycle C_k, $k \geq 4$, as an induced subgraph. There is a vast scientific literature on chordal graphs. Their fame is due to their classic and beautiful characterizations, various mathematical properties, and numerous applications in combinatorial optimization, relational databases, studies in phylogeny, Bayesian networks, record allocation, and solution of matrix systems, among other applications. For interested readers, we recommend the classic book by Martin C. Golumbic (1980), in which chordal graphs are called *triangulated graphs*.

In this section, we will prove the following result, one of the most fundamental in studies of convex geometries in graphs, proved by Farber and Jamison (1986): a graph G is chordal if and only if the monophonic convexity of G is a convex geometry. As seen in Sect. 2.3, the monophonic convexity of a graph G is defined over the *induced* paths of G.

We recall that, for a vertex $v \in V(G)$, the *closed neighborhood* of v is the set formed by v and its neighbors, denoted by $N[v]$. In addition, v is *simplicial* when $N[v]$ is a clique.

Suppose that S is a convex set in the monophonic convexity C of G, and v is an extreme vertex of S. Note that there is no induced path in $G[S]$ that contains v as an internal vertex. But this implies that v cannot have two nonadjacent neighbors in S, i.e., $N[v] \cap S$ is a clique. Therefore, v is a simplicial vertex in $G[S]$. On the other hand, it is easy to see that if v is a simplicial vertex in $G[S]$, then v is an extreme vertex of S. Therefore, $\text{Ext}(S) = \{v \mid v \text{ is simplicial in } G[S]\}$. Figure 4.4 illustrates this argument.

Theorem 4.1 (Farber and Jamison 1986) *A graph G is chordal if and only if the monophonic convexity of G is a convex geometry.*

Proof Suppose that G is a chordal graph. We will show that the monophonic convexity C of G is a convex geometry. A fundamental property of chordal graphs states that every non-simplicial vertex of G is in an induced path between two simplicial vertices (Farber and Jamison 1986). Therefore, every vertex of $V(G)$ is in

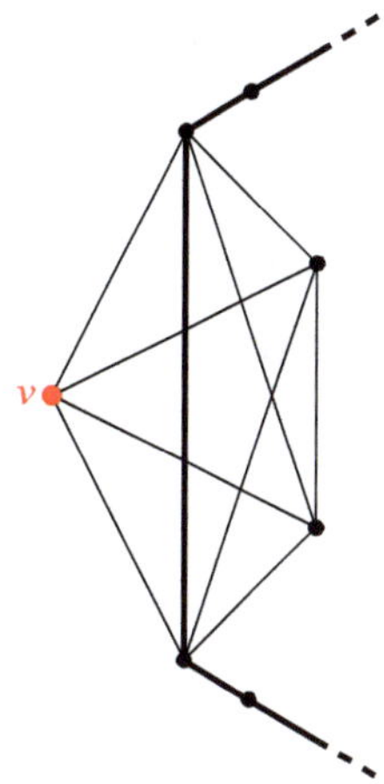

Fig. 4.4 In the figure, v is a simplicial vertex. The path marked with highlighted edges is induced and does not pass through v, but through its neighborhood. If any other path has v as an internal vertex, it must contain two neighbors of v and, therefore, will not be induced

an induced path between two extreme vertices of $V(G)$, which is trivially a convex set. In other words, $V(G) = \text{I}(\text{Ext}(G)$; but it is clear that $\text{I}(V(G)) = V(G)$, and, therefore, $V(G) = \text{conv}(\text{Ext}(G))$. We then conclude that $V(G)$ is the convex hull of its extreme vertices.

Now consider a convex set $S \subseteq V(G)$ in the monophonic convexity. Since the class of chordal graphs is hereditary (that is, every induced subgraph of a chordal graph is also a chordal graph), then $G[S]$ is a chordal graph. Using the arguments from the previous paragraph, we conclude likewise that $S = \text{conv}(\text{Ext}(S))$. Therefore, C satisfies the Minkowski–Krein–Milman property, i.e., C is a convex geometry.

Now suppose that C is a convex geometry. In order to obtain a contradiction, suppose that G contains an induced cycle C with at least four vertices. Note that the vertices of C are not simplicial vertices of G and, therefore, C does not contain extreme vertices of $V(G)$. Let $S = \text{conv}(C)$. Clearly, the vertices of C are not extreme vertices of S either. Furthermore, by the definition of convex hull, note that each vertex $x \in S \setminus V(C)$ was added to S precisely because it is an internal vertex of an induced path between two other vertices of S. Therefore, $x \notin \text{Ext}(S)$. We conclude that $\text{Ext}(S) = \emptyset$ and S is not the convex hull of its extreme vertices, i.e., the Minkowski–Krein–Milman property fails for S, a contradiction. □

4.3 Geodesic Convexity and Ptolemaic Graphs

A graph G is *Ptolemaic* if G does not contain the cycle C_k, $k \geq 4$, and the gem (Fig. 4.3) as induced subgraphs. Note that a Ptolemaic graph is chordal. Another important property establishes that Ptolemaic graphs are distance-hereditary: every induced path is a shortest path (Howorka 1981). In Fig. 4.3, the induced path $P = abcd$ between vertices a and d has length three. However, this path is not minimum, as the path $P' = aud$ has length two. Therefore, the gem is a forbidden structure for a Ptolemaic graph.

Suppose that S is a geodesically convex set. As in the previous section, it is simple to check that v is an extreme vertex of S if and only if v is simplicial (Exercise 4.3). With this information in hand, we can now prove the following result:

Theorem 4.2 (Farber and Jamison 1986) *A graph G is Ptolemaic if and only if the geodesic convexity of G is a convex geometry.*

Proof Suppose first that G is a Ptolemaic graph. As G is chordal, every non-simplicial vertex of G is in an induced path between two simplicial vertices. However, in a Ptolemaic graph, any induced path is a shortest path. Therefore, every vertex of $V(G)$ is in a shortest path between two extreme vertices of $V(G)$, i.e., $V(G)$ is the convex hull of its extreme vertices.

Since the class of Ptolemaic graphs is hereditary, then $G[S]$ is a Ptolemaic graph for any geodesically convex set $S \subseteq V(G)$ and, therefore, using the same arguments above, $S = \text{conv}(\text{Ext}(S))$. Then, the geodesic convexity of G is a convex geometry.

Now, suppose that the geodesic convexity of G is a convex geometry. As in the proof of Theorem 4.1, G cannot contain induced cycles with four or more vertices. In addition, considering the discussion presented in Example 4.1, it is easy to show that the convex hull of a gem in the geodesic convexity does not satisfy the Minkowski–Krein–Milman property (Exercise 4.4). Therefore, G cannot contain gems as induced subgraphs, which leads to the conclusion that G is Ptolemaic. □

4.4 Triangle-Path Convexity and Acyclic Graphs

In the two previous sections, in order to prove that a given convexity is a convex geometry, we directly used the Minkowski–Krein–Milman property. We can also use the anti-exchange property, an equivalent alternative. This is what we will do next.

Let $P = v_1 v_2 \dots v_k$ be a path in a graph G. A *chord* of P is any edge that connects nonconsecutive vertices of the path. A *triangle chord* is any edge of the form $v_i v_{i+2}$, with $1 \leq i \leq k-2$. The path P is called a *triangle path* if P admits only triangle chords. Figure 4.5 illustrates this concept.

In the triangle-path convexity, a set S is tp-*convex* if, for any $x, y \in S$, every vertex that belongs to some triangle path between x and y also belongs to S (Changat and Mathews 1999; Dourado and Sampaio 2016).

Theorem 4.3 (Dourado et al. 2025) *The triangle-path convexity of G is a convex geometry if and only if G is a forest (acyclic graph).*

Proof Suppose that the triangle-path convexity of G is a convex geometry. If there are vertices $a, b, c \in V(G)$ that induce a K_3, then $a \in \text{conv}(\{b, c\})$ and $b \in \text{conv}(\{a, c\})$, which contradicts the anti-exchange property, since $\{c\}$ is a tp-

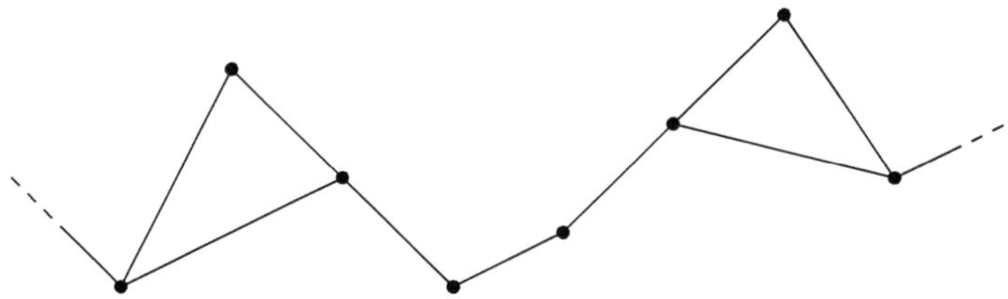

Fig. 4.5 A triangle path

convex set. Therefore, G does not contain K_3, and the extreme vertices of any edge of G form a tp-convex set. Furthermore, G does not contain any induced cycle C_k for $k \geq 4$, because if there is such a cycle $C_k = v_1 \dots v_k v_1$ in G, then $S = \{v_2, v_3\}$ is a tp-convex set, $v_1 \in \text{conv}(S \cup \{v_4\})$ and $v_4 \in \text{conv}(S \cup \{v_1\})$, which again contradicts the anti-exchange property. Therefore, G is an acyclic graph.

Conversely, suppose that G is an acyclic graph. Since G does not contain K_3 as a subgraph, the triangle-path convexity and the monophonic convexity of G are identical. In addition, G is clearly chordal, which implies, by Theorem 4.1, that such convexity is a convex geometry. □

A Note on the All-Path Convexity

A set S is ap-*convex* if, for any $x, y \in S$, every vertex that belongs to some path between x and y also belongs to S. The convexity formed by the ap-convex sets of a graph G is the *all-path convexity* of G.

It is easy to see that a vertex v is an extreme vertex of an ap-convex set S (with at least two vertices) if and only if v is a pendant vertex (or a *leaf*) in $G[S]$, because it is clear that v is an internal vertex of a path if and only if v has at least two neighbors.

If the all-path convexity of a graph G is a convex geometry, then G cannot contain cycles, because the degree of all vertices in a cycle C is at least two, and, therefore, $\text{conv}(V(C))$ cannot be the convex hull of its extreme vertices, since they do not exist. This discussion shows that G is acyclic.

On the other hand, if G is acyclic, then every non-leaf vertex is in a path between two leaves, and this property holds for any induced subgraph of G. Therefore, every ap-convex set of G is the convex hull of its extreme vertices, i.e., the all-path convexity of G is a convex geometry.

The arguments above prove the following theorem:

Theorem 4.4 *The all-path convexity of G is a convex geometry if and only if G is an acyclic graph.*

It is interesting to note, by Theorems 4.3 and 4.4, that the triangle-path and all-path convexities of an acyclic graph G coincide. However, they do not coincide in general. For example, in the cycle C_k with $k \geq 3$, the family of tp-convex sets of

C_k is different from the family of ap-convex sets of C_k, since edges are tp-convex but not ap-convex.

4.5 P_3 Convexity and Forests of Stars

A *star* is a graph formed by a vertex v of degree $k \geq 0$ (called the *central vertex*) and k vertices of degree one adjacent to v. A *forest of stars* is a graph such that each of its connected components is a star.

It can be shown that a graph G is a forest of stars if and only if the P_3 convexity of G is a convex geometry (Dourado et al. 2025). The proof of this result is left for Exercise 4.3. The hint is to use the techniques shown in previous sections: application of the Minkowski–Krein–Milman property or the anti-exchange property.

The P_3^* convexity is discussed in the next section, as it is equivalent to the l^k convexity, defined below, for $k = 2$.

4.6 l^2 Convexity and Chordal Cographs

A graph G is a *cograph* if G does not contain the graph P_4 as an induced subgraph. Likewise chordal graphs, there is a vast scientific literature on cographs, including characterizations, algorithms, and applications. Cographs admit a very ordered structure and, therefore, many computationally difficult problems in graph algorithm theory become tractable when restricted to cographs. A reference on cographs is the book by Brandstädt and Spinrad (1999).

A set S is l^k*-convex* if, for any $x, y \in S$, every vertex that belongs to some induced path of length at most k between x and y also belongs to S (Gutierrez et al. 2023). The convexity formed by the l^k-convex sets of a graph G is called the l^k *convexity* of G. Note that the l^2 convexity coincides with the P_3^* convexity.

A graph G is a *chordal cograph* if G is simultaneously a cograph and a chordal graph, i.e., it does not contain the graphs P_4 and C_k ($k \geq 4$) as induced subgraphs. It can be shown that a graph G is a chordal cograph if and only if the l^2 convexity of G is a convex geometry (Gutierrez et al. 2023). The proof is left for Exercise 4.6.

4.7 m^3 Convexity and Weakly Polarized Graphs

A graph G is *weakly polarized* if G does not contain the graph C_k ($k \geq 5$) nor any of the graphs of Fig. 4.6 as an induced subgraph.

Weakly bipolarized graphs were introduced by Olariu (1989). We will see in this section that G is a weakly bipolarized graph if and only if the m^3 convexity of G is a convex geometry (Dragan et al. 1999). A set S is m^3*-convex* if, for any $x, y \in S$,

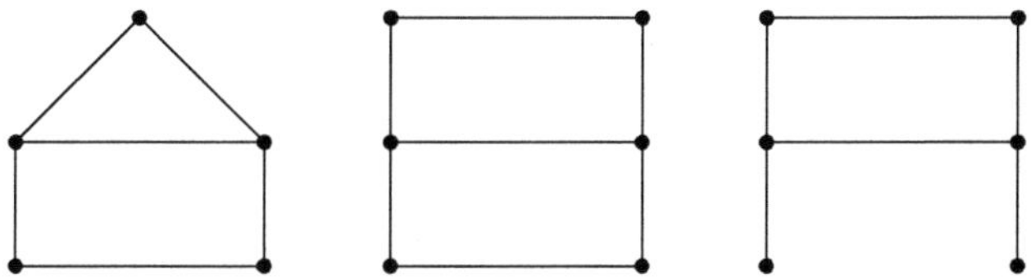

Fig. 4.6 From left to right: house, domino and "A" graphs

every vertex that belongs to some induced path of length at least three between x and y also belongs to S. Note that an m^3-convex set of a connected graph does not necessarily induce a connected subgraph (Exercise 4.7). The convexity formed by the m^3-convex sets of G is called the m^3 *convexity* of G.

An alternative definition of simplicial vertex is the following: v is a simplicial vertex if and only if v is not the central vertex of any induced path with three vertices (Exercise 4.8). In (Jamison and Olariu 1988), this concept is relaxed in the following way: a vertex v is *semi-simplicial* if v is not an internal vertex of any induced P_4 (an induced path with four vertices). Clearly, a vertex v is an extreme vertex of an m^3-connected set $S \subseteq V(G)$ if and only if v is semi-simplicial in $G[S]$.

We are now in a position to present a summarized proof of the following theorem:

Theorem 4.5 (Dragan et al. 1999) *A graph G is weakly bipolarized if and only if the m^3 convexity of G is a convex geometry.*

Proof Suppose that the m^3 convexity of G is a convex geometry and G contains a house G' as an induced subgraph (Fig. 4.6). Note that the only possible semi-simplicial vertex in the subgraph of G induced by $Y = \text{conv}(V(G'))$ is the vertex of degree two in the figure. Let x be this vertex. This implies that $\text{Ext}(Y) \subseteq \{x\}$ and, therefore, $\text{conv}(\text{Ext}(Y)) \neq Y$, a contradiction. Similarly, G cannot contain the cycle C_k ($k \geq 5$), the domino graph, and the graph A as induced subgraphs. Consequently, G is weakly bipolarized.

Conversely, if G is weakly bipolarized, it can be shown (Dragan et al. 1999) that every vertex v that is not semi-simplicial is in an induced path of length at least three between two semi-simplicial vertices. In other words, $V(G) = \text{I}(\text{Ext}(G)) = \text{conv}(\text{Ext}(G))$. Since weakly bipolarized graphs are hereditary, the same argument applies to any induced subgraph of G, in particular to the subgraphs induced by m^3-convex sets of G. Therefore, the m^3 convexity of G is a convex geometry. □

4.8 Strong Convexity and Strongly Chordal Graphs

We say that a path $P = u_0u_1 \ldots u_n$ in a graph G is *even-chorded* if, for every chord u_iu_j, $|i-j|$ is even and $\{i, j\} \cap \{0, n\} = \emptyset$. That is, the chords of P connect vertices that are at an even distance in P, and both vertices u_0 and u_n are in no chord of P. See Fig. 4.7.

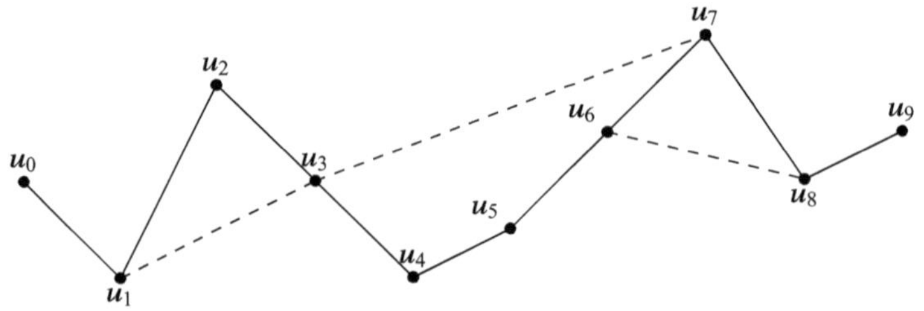

Fig. 4.7 An even-chorded path. Chords are marked with dotted lines

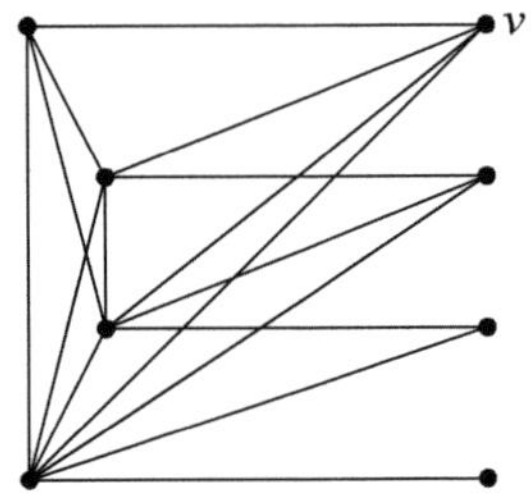

Fig. 4.8 In the graph above, v is a simple vertex

A set S is *strongly convex* if, for any $x, y \in S$, every vertex that belongs to some even-chorded path between x and y also belongs to S. The convexity formed by the strongly convex sets of a graph G is called the *strong convexity* of G.

An *even cycle* is a cycle with an even number of vertices. A graph G is *strongly chordal* (Farber 1983) if G is chordal and each of its even cycles with at least six vertices has an odd chord (a chord that connects two vertices that are at an odd distance in the cycle). See Exercise 4.9.

A vertex v of a graph G is *simple* if the closed neighborhoods of its neighbors form a nested set family, i.e., the set $N(v)$ admits an ordering $N(v) = \{v_1, \ldots, v_k\}$ such that $N[v_1] \subseteq N[v_2] \subseteq \cdots \subseteq N[v_k]$. See an example in Fig. 4.8. Note that a simple vertex is simplicial (Exercise 4.10), but not conversely.

Farber (1983) provided the following characterization of strongly chordal graphs:

Theorem 4.6 (Farber 1983) *A graph G is a strongly chordal graph if and only if every induced subgraph of G contains a simple vertex.*

The above theorem is the version for strongly chordal graphs of the well-known characterization of chordal graphs that says a graph G is chordal if and only if every induced subgraph of G contains a simplicial vertex—see, for example, the book by Golumbic (1980).

We can now state the main result of this subsection:

Theorem 4.7 (Farber and Jamison 1986) *A graph G is strongly chordal if and only if the strong convexity of G is a convex geometry.*

Proof Suppose that the strong convexity of G is a convex geometry, and consider $S \subseteq V(G)$. Let $S' = \text{conv}(S)$. Then, S' is strongly convex and $\text{Ext}(S') \subseteq S$. As every induced path is an even-chorded path, an extreme vertex v of S' is necessarily simplicial in $G[S']$ and therefore in $G[S]$. This implies that every induced subgraph of G has a simplicial vertex, i.e., G is chordal. It is not difficult to also prove that, as G is chordal, v is a simple vertex in $G[S']$ and therefore in $G[S]$ (Exercise 4.11). Hence, every induced subgraph of G has a simple vertex. By Theorem 4.6, G is strongly chordal.

Suppose now that G is strongly chordal. In (Farber and Jamison 1986), it is proved that every non-simple vertex of a strongly chordal graph is in an even-chorded path between two simple vertices of G. In other words, by Exercise 4.11, $V(G)$ is the convex hull of its extreme vertices. As the class of strongly chordal graphs is hereditary (Exercise 4.9), the same argument applies to any strong convex set S of G. Therefore, the strong convexity of G is a convex geometry. □

4.9 Toll Convexity and Interval Graphs

A walk $u_0u_1 \dots u_{k-1}u_k$ is a *toll walk* if $u_0u_i \in E(G)$ implies $i = 1$, and $u_ju_k \in E(G)$ implies $j = k-1$. In other words, u_1 is the only vertex of the walk adjacent to u_0 and, moreover, u_1 occurs only once in the walk (if $u_1 = u_j$ for $j \neq 1$, we would have $u_0u_j \in E(G)$, violating the definition of toll walk). The same observation applies to the pair u_k, u_{k-1}. The idea of this definition is as follows: the walk starts at u_0, passes through a *toll* (the vertex u_1) only once, and does not return to the toll. For all practical purposes, we are only interested in walks where the origin and destination are distinct vertices ($u_0 \neq u_k$). Let us make this assumption for the rest of the chapter.

Example 4.2 In the graph K_n, the only possible toll walk is of the form u_0u_k (a walk consisting of a single edge). In all other walks, the vertex u_0 (respectively the vertex u_k) will be adjacent to some internal vertex u_j with $j \neq 1$ (respectively $j \neq k-1$), since the graph is complete.

In (Alcón et al. 2015), toll walks were used to characterize interval graphs. A graph G is an *interval graph* if its vertices can be associated with intervals on the real line in such a way that two vertices are adjacent if and only if the intervals associated with them intersect. The intervals associated with the vertices of G form an *interval model* $\mathcal{I} = \{I_v\}_{v \in V(G)}$ of G. Figure 4.9 displays an example of an interval graph. We can assume without loss of generality that the endpoints of the intervals are all distinct from each other. Information about interval graphs, with applications, can be found in the book by Golumbic (1980).

We say that a set S is *toll-convex* if, for any $x, y \in S$, every vertex that belongs to some toll walk between x and y also belongs to S. The convexity formed by the toll-convex sets of a graph G is called the *toll convexity* of G. Given an interval I on the real line, let $R(I)$ and $L(I)$ be the right and left endpoints of I, respectively. Given an interval model $\mathcal{I} = \{I_v\}_{v \in V(G)}$, for an interval graph G, we say that I_a is a *terminal interval* if $L(I_a) = Min \bigcup I_v$ or $R(I_a) = Max \bigcup I_v$, i.e., I_a occurs as

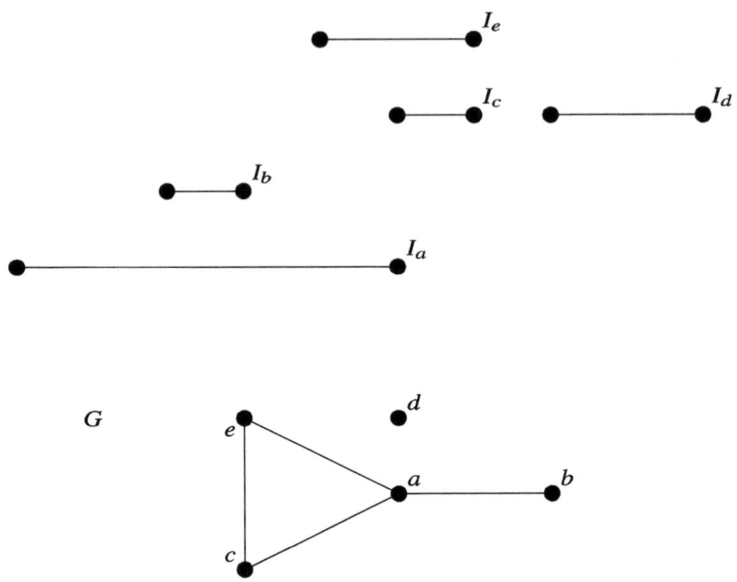

Fig. 4.9 An interval graph G (below) and an interval model of G (above)

the *first interval* in $\mathcal{J}$ (which contains the leftmost endpoint) or the *last interval* in $\mathcal{J}$ (which contains the rightmost endpoint). A vertex $a \in V(G)$ is a *terminal vertex* if there is an interval model of G in which a is associated with a terminal interval. Furthermore, the vertex a is a *terminal simplicial vertex* if a is both a terminal vertex and a simplicial vertex.

Example 4.3 Consider again Fig. 4.9. The terminal simplicial vertices of G are b, c, d, e (why?). Vertex a, although associated in Fig. 4.9 with a terminal interval, is not a simplicial vertex.

In (Alcón et al. 2015), three important facts about extreme vertices of toll-convex sets are provided:

Lemma 4.2 (Alcón et al. 2015)

(*a*) *If v is an extreme vertex of a toll-convex set S of G, then v is simplicial in $G[S]$.*
(*b*) *A vertex v of an interval graph G is an extreme vertex of a toll-convex set $S \subseteq V(G)$ if and only if v is a terminal simplicial vertex of $G[S]$.*
(*c*) *Let G be an interval graph. Then every vertex of G that is not terminal simplicial is in a toll walk between two terminal simplicial vertices.*

Example 4.4 As an example of Lemma 4.2(c), recall that in Fig. 4.9 vertex a is the only vertex that is not terminal simplicial in G. Moreover, a is an internal vertex of the walk $P = bac$, a toll walk with endpoints that are terminal simplicial vertices (b and c). Finally, note that each of the vertices b, c, d, e is not an internal vertex of any toll walk.

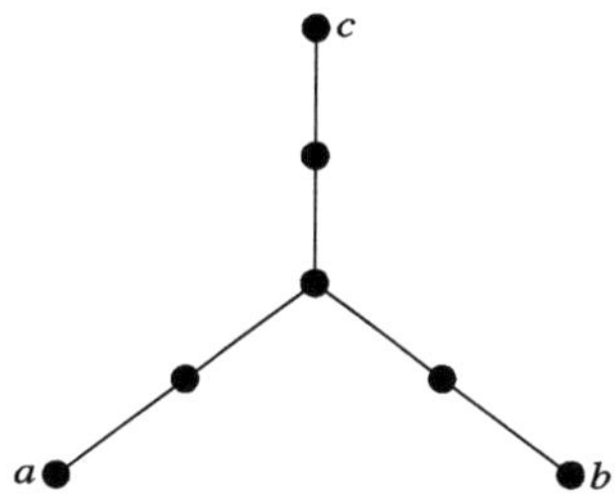

Fig. 4.10 Vertices a, b, c form an asteroidal triple

To characterize graphs with toll convexities that are convex geometries, we need to resort to an important characterization of interval graphs. Three distinct vertices of a graph form an *asteroidal triple* if between any two of them there is a path that avoids the neighborhood of the third. For example, in Fig. 4.10, vertices a, b, c form an asteroidal triple.

Theorem 4.8 (Lekkerkerker and Boland 1962) *A graph G is an interval graph if and only if G is chordal and does not contain asteroidal triples.*

We now state the main result of this section:

Theorem 4.9 (Alcón et al. 2015) *A graph G is an interval graph if and only if the toll convexity of G is a convex geometry.*

Proof Suppose that G is an interval graph. By Lemma 4.2, items (b) and (c), $V(G)$ is the convex hull of its extreme vertices. As interval graphs form a hereditary class (Lekkerkerker and Boland 1962), any toll-convex set $S \subseteq V(G)$ induces an interval graph, and the same argument applies again—S is the convex hull of its extreme vertices. Therefore, the toll convexity of G is a convex geometry.

Now suppose that the toll convexity of G is a convex geometry. We will prove that G is chordal. By hypothesis, every toll-convex set $S \subseteq V(G)$ is the convex hull of its extreme vertices. Let v be an extreme vertex of $V(G)$. By definition, $V(G) \setminus \{v\}$ is toll-convex in G. Therefore, the toll convexity of $G - v$ is a convex geometry, as any set $S' \subseteq V(G) \setminus \{v\}$ is toll-convex in $G - v$ if and only if S' is toll-convex in G. This implies, by induction, that $G - v$ is chordal; furthermore, by Lemma 4.2(a), v is a simplicial vertex of G. Therefore, G is also a chordal graph—recall that G is chordal if and only if every induced subgraph of G has a simplicial vertex (Golumbic 1980).

We will now show that G does not contain asteroidal triples. Suppose, for the sake of contradiction, that a, b, c form an asteroidal triple in G. Consider the walk W from a to c formed by the concatenation of induced paths P_{ab}, from a to b, and P_{bc}, from b to c, such that P_{ab} avoids $N[c]$ and P_{bc} avoids $N[a]$. Note that a is adjacent only to one vertex of P_{ab} and to no vertex of P_{bc}. Similarly, c is adjacent only to one vertex of P_{bc} and to no vertex of P_{ab}. Therefore, W is a toll walk from a to c passing through b. Let $Y = \text{conv}(\{a, b, c\})$. As $V(W) \subseteq Y$, b is not an extreme vertex of Y. Similarly, a and c are not extreme vertices of Y. But no other vertex

$x \in Y$ is an extreme vertex of Y. This implies that Y does not contain extreme vertices, a contradiction. Thus, by Theorem 4.8, G is an interval graph. □

4.10 Weakly Toll Convexity and Unit Interval Graphs

A walk $P = u_0u_1 \ldots u_{k-1}u_k$ is a *weakly toll walk* if $u_0u_i \in E(G)$ implies $u_i = u_1$ and $u_ju_k \in E(G)$ implies $u_j = u_{k-1}$. In other words, the only vertex adjacent to u_0 (respectively, u_k) in the walk P is u_1 (respectively, u_{k-1}), which may occur more than once in the walk.

As in the previous section, we assume in this section that $u_0 \neq u_k$.

Example 4.5 Note that every toll walk is also a weakly toll walk, but the reverse is not always true. Consider, for example, the graph $K_{1,3}$ with vertex set $\{a, b, c, d\}$, where the vertex of degree three is vertex a. Note that there is no toll walk containing vertex b as an internal vertex. However, there is a weakly toll walk containing vertex b as an internal vertex (for example, the walk $P = cabad$). The same applies to vertices c and d.

In the previous section, toll walks were used to characterize interval graphs through convex geometries. Similarly, weakly toll walks will now be used to characterize unit interval graphs.

A *unit interval graph* or *proper interval graph* is an interval graph that admits an interval model in which all intervals have the same length; or, equivalently, in which no interval is contained in another interval.

Roberts (1969) proved that unit interval graphs are exactly the interval graphs that do not contain $K_{1,3}$ as an induced subgraph.

Example 4.6 Consider again the graph $K_{1,3}$. It is easy to see that $K_{1,3}$ is an interval graph, but it is not a unit interval graph (Exercise 4.12).

We say that a set S is *wto-convex*[1] if, for any $x, y \in S$, every vertex that belongs to some weak toll walk between x and y also belongs to S. The convexity formed by the wto-convex sets of a graph G is called the *weak toll convexity* of G.

Lemma 4.2 has an analogous version for wto-convex sets:

Lemma 4.3 (Alcón et al. 2015)

(*a*) *If v is an extreme vertex of a wto-convex set S of a graph G, then v is simplicial of $G[S]$.*

(*b*) *A vertex v of a unit interval graph G is an extreme vertex of a wto-convex set $S \subseteq V(G)$ if and only if v is a terminal simplicial vertex of $G[S]$.*

(*c*) *Let G be a unit interval graph. Then every vertex of G that is not a terminal simplicial vertex is in a weak toll walk between two terminal simplicial vertices.*

---

[1] Here, we use the prefix "wto" for *weakly toll*.

Using arguments similar to those used in Theorem 4.9, we can prove:

Theorem 4.10 (Alcón et al. 2015) *A graph G is a unit interval graph if and only if the weak toll convexity of G is a convex geometry.*

4.11 Hereditary Versus Nonhereditary Graph Classes

As we have seen, the results of this chapter can be placed within a general scheme, like a *recipe* for producing results about characterizations of hereditary graph classes through convex geometries:

- We consider a hereditary graph class $\mathcal{G}$ that can be characterized by *forbidden induced subgraphs*. This means that there is a family $\mathcal{F}$ of graphs such that a graph G belongs to the class $\mathcal{G}$ if and only if G does not contain any graph in $\mathcal{F}$ as an induced subgraph.
- We consider an appropriate rule r for the definition of convex sets, based on some system $\mathcal{P}$ of paths or walks: a set S is r-convex if, for any $x, y \in S$, every vertex that belongs to some member of $\mathcal{P}$ between x and y also belongs to S.
- We consider the corresponding convexity C formed by r-convex sets.
- Given a graph $G \in \mathcal{G}$, we characterize the extreme vertices of r-convex sets of the convexity C of G.
- We prove the following theorem: *G belongs to the class $\mathcal{G}$ if and only if the convexity C of G is a convex geometry.*
- For the proof of necessity, we initially show that every non-extreme vertex of G is in a member of $\mathcal{P}$ between two extreme vertices. Next, we use the heredity of $G \in \mathcal{G}$ to show that this holds for any induced subgraph of a convex set of C.
- For the proof of sufficiency, we show that if there is an induced subgraph G' of G such that $G' \in \mathcal{F}$, then $S = \text{conv}(V(G')) \neq \text{conv}(\text{Ext}(S))$. This argument directly uses the definition of the Minkowski–Krein–Milman property, but we can alternatively use the anti-exchange property.

An interesting exercise is to review the preceding sections and locate in each of them the elements of the above recipe (Exercise 4.13).

Another less elegant recipe to characterize a hereditary class $\mathcal{G}$ of graphs through convex geometries is to work with convexities not defined on systems of paths or walks. Let G and H be graphs, with $|V(H)| \geq 2$. Remember that, in Sect. 2.4, we defined a subset $S \subseteq V(G)$ as H-free convex if no vertex outside S induces H with some subset of S. In addition, as seen in Lemma 2.6, the family of H-free convex sets form a convexity, called H-free convexity. Given a family $\mathcal{F}$ of graphs with at least 2 vertices, we say that $S \subseteq V(G)$ is $\mathcal{F}$-free convex if S is H-free convex for every $H \in \mathcal{F}$. We call the convexity associated with the $\mathcal{F}$-free convex sets *$\mathcal{F}$-free convexity*. We say that G is $\mathcal{F}$-free if G does not contain any graph in $\mathcal{F}$ as an induced subgraph.

Theorem 4.11 (Dourado et al. 2025) *Let $\mathcal{G}$ be a hereditary graph class and let $\mathcal{F}$ be the minimal family of graphs such that $G \in \mathcal{G}$ if and only if G is $\mathcal{F}$-free. Therefore, a graph G is a member of $\mathcal{G}$ if and only if the $\mathcal{F}$-free convexity of G is a convex geometry.*

Proof First, suppose that $G \in \mathcal{G}$, i.e., G is $\mathcal{F}$-free. Then, by definition, every subset of $V(G)$ is $\mathcal{F}$-free convex, meaning that the anti-exchange property is valid. Therefore, the $\mathcal{F}$-free convexity of G is a convex geometry. On the other hand, suppose that the $\mathcal{F}$-free convexity of G is a convex geometry and that G contains a graph of $\mathcal{F}$ as an induced subgraph. Let $S \subseteq V(G)$ be a minimum set such that $G[S]$ is isomorphic to a graph H of $\mathcal{F}$. Let x, y be distinct vertices of S and let $S' = S \setminus \{x, y\}$. By the minimality of S, note that S' is $\mathcal{F}$-free convex. Also note that $x \in \text{conv}(S' \cup \{y\})$ and $y \in \text{conv}(S' \cup \{x\})$, meaning that the anti-exchange property is not valid. Contradiction, since the $\mathcal{F}$-free convexity of G is a convex geometry. Therefore, G is $\mathcal{F}$-free. □

With this emphasis on hereditary classes, we now pose the following question. Is there any *nonhereditary* graph class that can be characterized via convex geometries and path systems? The answer is yes, and we will see why in the next section.

An Example of a Nonhereditary Class

As seen in Sect. 4.6, a set S is *l^k-convex* if, for any $x, y \in S$, every vertex in some induced path of length at most k between x and y also belongs to S. The convexity formed by the l^k-convex sets of a graph G is called the l^k *convexity* of G.

In (Gutierrez et al. 2023), a characterization of a graph class $\mathcal{G}$ is presented such that the l^3 convexity of every $G \in \mathcal{G}$ is a convex geometry. To understand its statement, we need some definitions.

An *n–gem* G_n ($n \geq 4$) is a graph with vertices $x_0, \ldots, x_n, u_n$ such that $x_0, \ldots, x_n$ induce a path and u_n is a universal vertex. See Fig. 4.11. Assume that G_n is an induced subgraph of a graph G. We say that G_n is *protected* if there exists in the graph G a P_4 between x_0 and x_n that does not contain u_n.

Observe in Fig. 4.11 that the path $P = u_0 \ldots u_n$ contains at least five vertices. Therefore, in the l^3 convexity, the convex hull of the set $\{u_0, u_n\}$ does not contain all the vertices of G_n. If the l^3 convexity of a graph G is a convex geometry, G cannot contain G_n as an induced subgraph, unless an additional condition allows this possibility; this condition is precisely that every n-gem of G is protected (condition (c) of the following theorem).

Theorem 4.12 (Gutierrez et al. 2023) *The l^3 convexity of G is a convex geometry if and only if the following conditions are satisfied:* (a) *G is chordal,* (b) *diam$(G) \leq 3$, and* (c) *every induced n-gem ($n \geq 4$) of G is protected.*

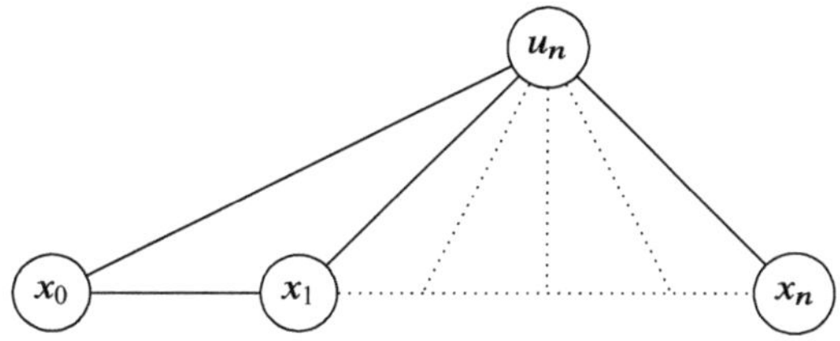

Fig. 4.11 The n-gem graph ($n \geq 4$)

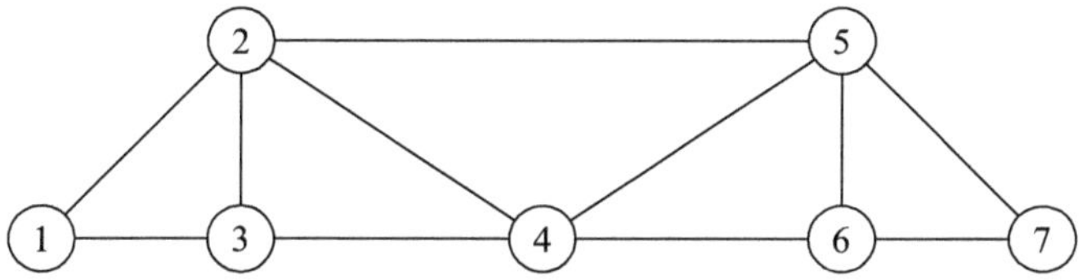

Fig. 4.12 The l^3 convexity of G is a convex geometry, but this property is not true for all induced subgraphs of G. This implies that the graph class $\mathcal{G} = \{G \mid$ the convexity l^3 of G is a convex geometry$\}$ is not hereditary

Consider the graph class $\mathcal{G} = \{G \mid$ the convexity l^3 of G is a convex geometry$\}$. We will show that, unlike the other graph classes presented in this chapter, $\mathcal{G}$ is not hereditary.

Example 4.7 Consider the graph G of Fig. 4.12. Note that $\text{Ext}(G) = \{1, 7\}$ and $\text{conv}(\{1, 7\}) = V(G)$. Therefore, in the l^3 convexity, $V(G)$ is the convex hull of its extreme vertices. It is not difficult to check (Exercise 4.14) that, for every l^3-convex set S of G, S is the convex hull of its extreme vertices, i.e., the l^3 convexity of G is indeed a convex geometry. However, for $x \in \{2, 5\}$, the l^3 convexity of the graph $G - x$ is not a convex geometry, because:

- $V(G - x)$ is trivially an l^3-convex set of $G - x$.
- $\text{Ext}(V(G - x)) = \{1, 7\}$.
- $\text{conv}(\{1, 7\}) = \{1, 7\} \neq V(G - x)$.

Exercises

Exercise 4.1 Prove Lemma 4.1: a convexity C is a convex geometry if and only if it satisfies the anti-exchange property.

Exercise 4.2 Prove that, in the successive application of the interval operator described in Sect. 4.2, the set S_k is equal to $\text{conv}_C(S)$.

Exercise 4.3 Prove that, in the geodesic convexity of a graph G, v is an extreme vertex of a convex set S if and only if v is simplicial in $G[S]$.

Exercise 4.4 Let H be an induced subgraph of a graph G such that H is isomorphic to a gem. Considering the geodesic convexity of G, show that $\text{conv}(H)$ is not the convex hull of its extreme vertices.

Exercise 4.5 Prove that a graph G is a forest of stars if and only if the P_3 convexity of G is a convex geometry.

Exercise 4.6 Prove that a graph G is a chordal cograph if and only if the l^2 convexity of G is a convex geometry.

Exercise 4.7 Show that an m^3-convex set of a connected graph does not necessarily induce a connected subgraph.

Exercise 4.8 Show that a vertex v is simplicial if and only if v is not the central vertex of any induced path with three vertices.

Exercise 4.9 Draw all possible configurations of cycles of six vertices with k chords, for $k \in \{0, \ldots, 9\}$. (Use symmetries to avoid drawing the same configuration more than once.) For each configuration, decide whether a strongly chordal graph can or cannot contain that configuration as an induced subgraph. Conclude that the class of strongly chordal graphs is hereditary by induced subgraphs.

Exercise 4.10 Show that every simple vertex is simplicial, and that the converse is not true.

Exercise 4.11 Let G be a chordal graph, and consider the strong convexity of G. Prove that v is an extreme vertex of a strong convex set S if and only if v is a simple vertex in $G[S]$.

Exercise 4.12 Show that the graph $K_{1,3}$ is an interval graph, but not a unit interval graph.

Exercise 4.13 Review Sects. 4.2 to 4.10, and try to locate in each of them all the elements described in the recipe presented in Sect. 4.11.

Exercise 4.14 List all the l^3-convex sets of the graph G in Fig. 4.12, and show that the l^3 convexity of G is a convex geometry.

Part II
Main Convexities and Applications

Chapter 5
P_3 and P_3^* Convexities

In this chapter, we study the convexities P_3 and P_3^* in more depth. Despite the geodesic convexity being the most traditional, many results of the P_3 and P_3^* convexities are used to obtain results in the geodesic convexity, mainly related to computational complexity and, therefore, we decided to present them first.

Remember that the convexity P_3 of a graph G is the convexity of paths P_3 of G. That is, a set of vertices $S \subseteq V(G)$ is P_3-convex if and only if no vertex $u \notin S$ has two neighbors in S. An interesting point is that S is an interval set in the P_3 convexity if and only if S is a 2-dominant set of G, where S is *2-dominant* if every vertex $u \in V(G) \setminus S$ has at least 2 neighbors in S. The concept of domination and its variants have many applications and constitute a widespread area of graph theory (Haynes et al. 2020).

Note that the hull number in the P_3 convexity can be used to model the propagation of influence in social networks (Kempe et al. 2003; Dreyer and Roberts 2009; Chen 2009). In this diffusion model in graphs, we say that the vertices under a given influence are *active* (or influenced) and active vertices remain active forever. An inactive vertex becomes active if it has two active neighbors. A natural problem is to obtain the smallest number of *influencers* (the set of vertices active at the beginning) in order to influence the entire graph after the diffusion process. This is exactly the hull number of the convexity P_3. A more general model is shown in Sect. 9.1.

In the P_3^* convexity, the diffusion is slightly different: a vertex is influenced if it has two active neighbors that are not adjacent to each other. For example, if someone receives a restaurant recommendation from two friends of different groups (family and university, for example), they are likely to take the recommendation more seriously. Formally speaking, the convexity P_3^* of a graph G is the convexity of induced paths P_3 of G. That is, a set of vertices $S \subseteq V(G)$ is P_3^*-convex if and only if no vertex $u \notin S$ has two neighbors in S that are not adjacent to each other.

J. Araújo et al., *Introduction to Graph Convexity*, SBMAC Collection on Applied and Computational Mathematics, https://doi.org/10.1007/978-3-031-84128-6_5

Note that $I_{p3*}(S) \subseteq I_{p3}(S)$. This implies that $\text{hn}_{p3}(G) \leq \text{hn}_{p3*}(G)$ and $\text{in}_{p3}(G) \leq \text{in}_{p3*}(G)$. However, it does not imply that $\text{con}_{p3}(G) \geq \text{con}_{p3*}(G)$, given the need to be a proper convex set. For example, if G' is the graph obtained from the graph with $n-2$ vertices and no edge, adding 2 universal vertices, then $\text{con}_{p3}(G') = 1$ and $\text{con}_{p3*}(G') = n-1$ (Exercise 5.1).

Convexity P_3 was introduced by Centeno et al. (2009) and convexity P_3^* was introduced by Araújo et al. (2013). As seen in Lemma 2.3, convexities P_3 and P_3^* are equivalent in triangle-free graphs, like bipartite graphs, because every P_3 in the graph is induced. With this, computational complexity results in the P_3 convexity for triangle-free graphs lead to the same results in the P_3^* convexity. For example, Barbosa et al. (2012) proved that the Carathéodory number of convexity P_3 has a polynomial algorithm for trees and, therefore, the same holds in the P_3^* convexity.

The theorem below summarizes complexity results for all parameters of the P_3 convexity in bipartite graphs.

Theorem 5.1 (Various Authors) *Regarding bipartite graphs in the P_3 convexity (and therefore also in P_3^*), all the 10 parameters are* NP-*hard:*

- *Hull and convexity numbers (Araújo et al. 2013)*
- *Interval number (Centeno et al. 2009)*
- *Carathéodory number (Barbosa et al. 2012)*
- *Radon number (Dourado et al. 2013a)*
- *Helly number (Dourado and da Silva 2017)*
- *Percolation time (Marcilon and Sampaio 2018b)*
- *Iteration time (Araújo et al. 2025)*
- *General position number (Yannakakis 1981)*
- *Rank (Ramos et al. 2014)*

5.1 Relationship Between P_3^* and Geodesic Convexities

Lemma 2.4 shows a useful reduction from the P_3^* convexity to the geodesic convexity, which works well because it obtains a graph with diameter two (all shortest paths have length 1 or 2, i.e., P_2 or P_3). The theorem below shows that this reduction leads to computational complexity results for all the 10 parameters of the geodesic convexity.

Theorem 5.2 (Various Authors) *If G is not complete and G_u is obtained from G by including ℓ universal vertices $u_1, \ldots, u_\ell$, then:*

1. $\text{hn}_g(G_u) = \text{hn}_{p3*}(G_u) = \text{hn}_{p3*}(G)$
2. $\text{in}_g(G_u) = \text{in}_{p3*}(G_u) = \text{in}_{p3*}(G)$
3. $\text{con}_g(G_u) = \text{con}_{p3*}(G_u) = \text{con}_{p3*}(G) + \ell$
4. $\text{ti}_g(G_u) = \text{ti}_{p3*}(G_u) = \max\{\text{ti}_{p3*}(G),\ 1\}$
5. $\text{tp}_g(G_u) = \text{tp}_{p3*}(G_u) = \max\{\text{tp}_{p3*}(G), 1\}$

6. $\text{cth}_\text{g}(G_u) = \text{cth}_{\text{p3}*}(G_u) = \max\{\text{cth}_{\text{p3}*}(G), 2\}$
7. $\text{rd}_\text{g}(G_u) = \text{rd}_{\text{p3}*}(G_u) = \omega(G) + \ell$
8. $\text{h}\ell_\text{g}(G_u) = \text{h}\ell_{\text{p3}*}(G_u) = \omega(G) + \ell$
9. $\text{gp}_\text{g}(G_u) = \text{gp}_{\text{p3}*}(G_u) = \max\{\text{gp}_{\text{p3}*}(G),\ \omega(G) + \ell\}$
10. $\text{rk}_\text{g}(G_u) = \text{rk}_{\text{p3}*}(G_u) = \max\{\text{rk}_{\text{p3}*}(G),\ \omega(G) + \ell\}$

where $\omega(G)$ is the size of the largest clique of G.

Proof The proof for $\text{ti}(\cdot)$ and $\text{gp}(\cdot)$ is explicitly found in (Araújo et al. 2025). The proof for the others, with the exception of $\text{h}\ell(\cdot)$ and $\text{rk}(\cdot)$, is found in (Araújo et al. 2013). Below we provide a general idea for the proof, left as an exercise.

We write $\text{I}^u(\cdot)$ and $\text{conv}^u(\cdot)$ to differentiate the interval functions and the convex hull of G_u from those of G. First, note that, if $S \subseteq V(G)$ has at least two nonadjacent vertices, then

$$\text{I}^u_\text{g}(S) = \text{I}^u_{\text{p3}*}(S) = \text{I}_{\text{p3}*}(S) \cup \{u_1, \ldots, u_\ell\},$$

$$\text{conv}^u_\text{g}(S) = \text{conv}^u_{\text{p3}*}(S) = \text{conv}_{\text{p3}*}(S) \cup \{u_1, \ldots, u_\ell\}.$$

For this, note that every minimum path in G_u between vertices of G is an edge or induced P_3. Also note that every induced P_3 in G is a minimum path in G_u. Finally, as S has two nonadjacent vertices, then $I(S)$ contains $u_1, \ldots, u_\ell$ in the P_3^* and geodesic convexities. The rest of the proof is left as an exercise, basically a case analysis for each parameter.

An interesting case is the Helly number. Once proving that $\text{rd}_\text{g}(G_u) = \omega(G)+\ell$, we have directly that $\text{h}\ell_\text{g}(G_u) = \omega(G) + \ell$, since, as seen in Chap. 3, every clique is Helly independent in the geodesic convexity and $\text{h}\ell(G) \le \text{rd}(G)$ by Levi (1951) inequality (see Theorem 3.3). □

With this, we obtain the theorem below regarding all the 10 parameters of the P_3^* and geodesic convexities.

Theorem 5.3 *In the P_3^* and geodesic convexities, the following ten parameters are* NP-*hard in graphs with diameter 2: the hull, interval, and convexity numbers; the Carathéodory, Radon, and Helly numbers; the iteration and percolation times; the rank; and the general position number.*

Proof Taking $\ell = n$ in Theorem 5.2, note that $\text{gp}_\text{g}(G_u) = \text{rk}_\text{g}(G_u) = \omega(G) + n$. With this, we obtain the result by Theorem 5.1, by Theorem 5.2, and by the fact that $\omega(G)$ is NP-hard. □

Similarly, it is possible to obtain strong inapproximability results for several parameters of the geodesic convexity from results of the P_3^* convexity, such as the following, which are valid for both P_3^* and geodesic:

- Hull number is APX-hard (Coelho et al. 2015)
- Interval number is $(1 - \varepsilon) \ln n$-inapproximable (Coelho et al. 2015)

- Carathéodory, Radon, and convexity numbers are $n^{1-\varepsilon}$-inapproximable (Coelho et al. 2015)
- Helly number is $n^{1-\varepsilon}$-inapproximable (Dourado and da Silva 2017)
- Percolation time is $(5/4\text{-}\varepsilon)$-inapproximable (Marcilon and Sampaio 2018b)

for any $\varepsilon > 0$, where a parameter is $f(n)$-inapproximable when it does not have a polynomial algorithm with approximation factor $f(n)$, unless P=NP (see Appendix B).

Note that, in the above list of inapproximability results, only the iteration time, the general position number, and the rank are not included. To finish the section, we show another simple reduction that obtains a strong inapproximability result for the rank in the P_3^* and geodesic convexities. As above, the reduction works because it obtains a graph with diameter two.

Theorem 5.4 *If G is not complete and G'_u is obtained from G by including two vertices u_1 and u_2 adjacent to every vertex of G and not adjacent to each other, then* $\mathrm{rk}_g(G'_u) = \mathrm{rk}_{p3*}(G'_u) = \omega(G)+1$. *Therefore, the rank of the P_3^* and geodesic convexities are $n^{1-\varepsilon}$-inapproximable for all $\varepsilon > 0$.*

Proof We can assume that $\mathrm{rk}_g(G'_u) \geq 3$. Let S be a convexly independent set of G'_u in the P_3^* convexity with $|S| \geq 3$. If S contains two vertices x and y not adjacent to each other, then $u_1, u_2 \in \mathrm{conv}(S \setminus \{s\})$ and consequently $V(G'_u) \in \mathrm{conv}(S \setminus \{s\})$ for all $s \in S \setminus \{x, y\}$, leading to a contradiction, since S is convexly independent. Therefore, S is a clique. Hence, $\mathrm{rk}_g(G'_u) = \omega(G) + 1$ and the result follows from the fact that $\omega(G)$ is $n^{1-\varepsilon}$-inapproximable for any $\varepsilon > 0$ (Zuckerman 2006). □

5.2 Results for Some Graph Classes

In this section, we cite some more results from the literature on the P_3 and P_3^* convexities. In the P_3^* convexity, besides the results of Theorem 5.1, obtained from the P_3 convexity, direct results were also obtained for graphs with triangles. For example, Dourado et al. (2022a) proved that the hull number is NP-hard in chordal graphs (which can contain many triangles) and obtained a linear algorithm for unit interval graphs in the P_3^* convexity (which can also contain many triangles).

The following results hold for the P_3 convexity and, in the case of triangle-free graphs, as already mentioned, also hold for the P_3^* convexity. Campos et al. (2015) obtained polynomial algorithms for several parameters in graphs with few P_4's. For the Carathéodory number, Coelho et al. (2014) obtained a polynomial algorithm in chordal graphs.

Nichterlein et al. (2013) proved that the hull number is W[2]-hard even in bipartite graphs.[1] Penso et al. (2015) proved that the interval number is W[2]-hard

[1] This proof is presented in Sect. 5.3.

and the interval and hull numbers are NP-hard even in planar graphs with maximum degree 4. Cappelle et al. (2022) proved that the interval number is NP-hard even in split graphs with diameter 2.

Benevides et al. (2015) proved that the percolation time is polynomial in chordal graphs. They also proved that it is NP-hard even in planar graphs and it is NP-complete to decide whether the percolation time is greater than or equal to k in general graphs for any fixed $k \geq 4$.[2] Marcilon and Sampaio (2018b) proved that it is polynomial to decide whether the percolation time is greater than or equal to k for any $0 \leq k \leq 3$, resolving the question of complexity for fixed k. In bipartite graphs, Marcilon and Sampaio (2018b) proved that it is NP-complete to decide whether the percolation time is greater than or equal to k for every fixed $k \geq 5$. They also proved that it is polynomial to decide whether it is greater than or equal to k for any $0 \leq k \leq 4$, resolving the complexity issue for fixed k in bipartite graphs. Regarding parameterized complexity, Marcilon and Sampaio (2018c) proved that the percolation time is W[1]-hard when parameterized by treewidth. Marcilon and Sampaio (2018a) also proved that, in grid graphs of maximum degree 3, the percolation time is NP-hard, but it is polynomial if the grid is solid (without holes).

5.3 Hull Number Is W[2]-Hard

A famous problem known to be W[2]-hard is the problem of finding a hitting set parameterized by the solution value (Downey and Fellows 2012).

Parameterized Hitting Set

Instance:	A set U, a family $\mathcal{F} \subseteq 2^U$ and an integer $k \geq 0$.
Parameter:	k.
Question:	Is there a $U' \subseteq U$ such that $\|U'\| \leq k$ and $U' \cap F \neq \emptyset$, for every $F \in \mathcal{F}$?

Given an instance $(U, \mathcal{F}, k)$ of the **Parameterized Hitting Set** problem, we can define its *incidence graph* $G(U, \mathcal{F}, k) = (V_U, V_{\mathcal{F}}, E)$ as a bipartite graph in which, for each element $u \in U$, there is a vertex v_u in V_U, for each subset $F \in \mathcal{F}$ there is an element $v_F \in V_{\mathcal{F}}$, and there is an edge $v_u v_F \in E$ if and only if $u \in F$ for each $u \in U$ and $F \in \mathcal{F}$. Moreover, a subset $U' \subseteq U$ such that $U' \cap F \neq \emptyset$, for every $F \in \mathcal{F}$, is said to be a *hitting set* of $\mathcal{F}$.

[2] This proof is presented in Sect. 5.4.

Next, we present a parameterized reduction to show the computational hardness of determining the parameter $\text{hn}_{\text{p3}}(G)$.

Parameterized P_3 Hull Number

Instance:	A graph G and an integer $k \geq 0$.
Parameter:	k.
Question:	Is $\text{hn}_{\text{p3}}(G) \leq k$?

Theorem 5.5 (Nichterlein et al. 2013) *The problem* ***Parameterized*** P_3 ***Hull Number*** *is* W[2]*-hard, even when restricted to bipartite graphs of diameter 4.*

Proof Given an instance $I = (U, \mathcal{F}, k)$ of **Parameterized Hitting Set**, we construct an equivalent instance $I' = (G, k + 2)$ of **Parameterized P_3 Hull Number** in polynomial time in $n + m$, where $n = |U|$ and $m = |\mathcal{F}|$. In the following, note that the parameter of I' is $k + 2 = f(k)$, where k is the parameter of I. Therefore, such a reduction is indeed parameterized. Assume that $\mathcal{F} = \{F_1, \ldots, F_m\}$. For the construction of G, we first obtain the incidence graph $G(I)$, defined above, and add three vertices x, x_1, and x_2, along with the edges x_1x and x_2x. At the end of the construction, the vertices x_1 and x_2 will maintain degree 1 and, therefore, should belong to every P_3-hull set of G, as seen in Chap. 3. Also note that $x \in \text{I}_{\text{p3}}(\{x_1, x_2\})$, this vertex x being extremely useful in the construction, as well as serving in the argument that $\text{diam}(G) \leq 4$.

Initially, we add all edges xv_F for each $F \in \mathcal{F}$. Therefore, if any vertex b adjacent to v_F in G belongs to the hull $\text{conv}_{\text{p3}}(S)$ of a given subset S of vertices of G that contains x_1 and x_2, then v_F also belong to $\text{conv}_{\text{p3}}(S)$, since $x \in \text{I}_{\text{p3}}(\{x_1, x_2\})$.

Roughly speaking, we replace each vertex $v_u \in V_U$ of $G(I)$ by a path of C_4's. See Fig. 5.1. Formally let $J_u = \{j \in \{1, \ldots, m\} \mid u \in F_j\} = \{j_u^1, \ldots, j_u^{s_u}\}$, where $s_u = |J_u|$ is the number of subsets of $\mathcal{F}$ that have u as an element. Let G_u be the graph with vertices in $\{b_u^{s_u}\} \cup \bigcup_{i=1}^{s_u-1} \{a_u^i, b_u^i, c_u^i\}$ and edges in $\bigcup_{i=1}^{s_u-1} \{a_u^i b_u^i, a_u^i b_u^{i+i}, c_u^i b_u^i, c_u^i b_u^{i+i}\}$. Also add the edges $a_u^i x$ and $c_u^i x$ for each $i \in \{1, \ldots, s_u - 1\}$.

For each $j \in J_u$, the edge $v_u v_{F_j}$ of $G(i)$ is represented by the edge $b_u^i v_{F_j}$, which is added to G. Also add the edge $b_u^0 v_{F_{j_u^1}}$. Finally, also add a path $P_u = (p_u^1, q_u^1, \ldots, p_u^{s_u-1}, q_u^{s_u-1})$, along with the edges $q_u^i x$, for each $i \in \{1, \ldots, s_u - 1\}$, the edge $q_u^{s_u-1} b_u^1$, the edge $v_{F_{j_u^1}} p_u^1$ and the edges $v_{F_{j_u^i}} p_u^{i-1}$ for all $i \in \{2, \ldots, s_u\}$.

By replacing, for each $u \in U$, the vertex v_u with the graph G_u and adding the path P_u and the edges mentioned above, the construction of G is completed. Note that the construction of G can be obtained in polynomial time. See Exercise 5.5. Also note that G is bipartite. The vertices filled in gray in Fig. 5.1 help to observe

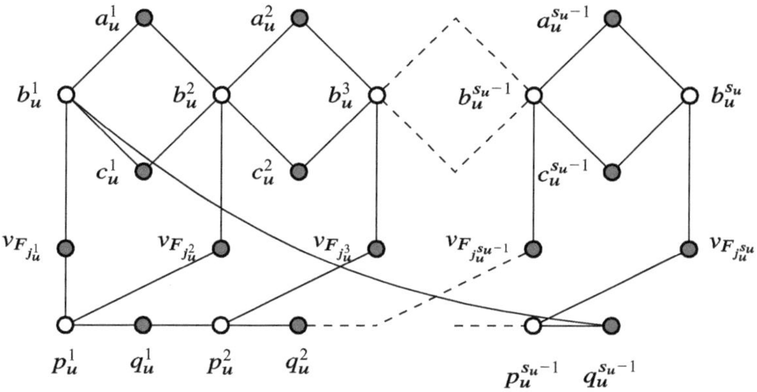

Fig. 5.1 Representation of the subgraph of G associated with an element $u \in U$. Vertices with gray interior have an edge to x

the bipartition of G. In addition, as previously stated, G has a diameter of 4, since every vertex represented in gray has an edge to x.

Now let us prove that $I = (U, \mathcal{F}, k)$ is a positive instance for **Parameterized Hitting Set** if and only if $\text{in}_{\text{p3}}(G) \leq k + 2$. Let $U' \subseteq U$ such that $|U'| \leq k$ and $U' \cap F \neq \emptyset$ for each $F \in \mathcal{F}$. Define $S = \{b_u^0 \mid u \in U'\} \cup \{x_1, x_2\}$. We claim that S is a P_3 hull set of G. In fact, it should be noted that $V(G_u) \subseteq \text{conv}_{\text{p3}}(\{b_u^0, x_1, x_2\})$ for each $u \in U$. Therefore, the vertices of the subgraphs G_u, when $u \in U'$, will belong to $\text{conv}_{\text{p3}}(S)$. Consequently, as U' is a transversal set of $\mathcal{F}$, we also have that all vertices v_F will belong to $\text{conv}_{\text{p3}}(S)$ for all $F \in \mathcal{F}$. This implies that, for every $u \in U$, all vertices of the path P_u will belong to $\text{conv}_{\text{p3}}(S)$. As $q_u^{s_u-1}$ and $v_{F_{j_u^1}}$ belong to $\text{conv}_{\text{p3}}(S)$, for every $u \in U$, we deduce that S is a hull set of G.

Now suppose that $\text{in}_{\text{p3}}(G) \leq k + 2$. Let S be a minimum hull set of G such that $|S| \leq k + 2$. As x_1 and x_2 have degree 1, $x_1, x_2 \in S$. As $x \in \text{I}_{\text{p3}}(\{x_1, x_2\})$, we know that $x \notin S$. The attentive reader may observe that we can assume that the remaining k vertices of S belong to the set $B = \{b_u^i \mid u \in U, i \in \{1, \ldots, s_u\}\}$, since a vertex r not belonging to B can be replaced by a vertex b from B, so that $r \in \text{conv}_{\text{p3}}((S \setminus \{r\}) \cup \{b\})$.

With this hypothesis, we claim that the set

$$U' = \{u \in U \mid \exists i \in \{1, \ldots, s_u\}(b_u^i \in S)\}\}$$

is a transversal set of $\mathcal{F}$. Note that the only vertex of G_u that has two neighbors in $V(G) \setminus V(G_u)$ is b_u^1. Therefore, if there is no vertex of G_u in S, since S is a hull set in the P_3 convexity, it is necessary that, in the iterative process of constructing $\text{conv}_{\text{p3}}(S)$, $q_u^{s_u-1}$ is added to $\text{conv}_{\text{p3}}(S)$ before any vertex of G_u. Note that this will only occur if all vertices v_{F_j} with $j \in J_u$ are already in $\text{conv}_{\text{p3}}(S)$. Therefore, the vertices chosen in S must be such that, in the iterative construction of $\text{conv}_{\text{p3}}(S)$, all vertices v_F are included in $\text{conv}_{\text{p3}}(S)$ before any vertex of a subgraph G_u that does not have any vertex from S is. This will only occur, therefore, if U' is a hull set in the P_3 convexity of G. □

5.4 Percolation Time Is $(4/3 - \varepsilon)$-Inapproximable

In this section, we show that the percolation times in the P_3 and P_3^* convexities are $(5/4 - \varepsilon)$-inapproximable for all $\varepsilon > 0$. That is, they do not have polynomial algorithms with that approximation factor, unless P=NP (see Appendix B).

Benevides et al. (2015) prove in Theorem 5.6 below that, in non-bipartite graphs, it is difficult to decide whether the P_3 percolation time is greater than or equal to 4. Marcilon and Sampaio (2018b) prove in Theorem 5.7 below that, in bipartite graphs, it is difficult to decide whether the P_3 (and P_3^*) percolation time is greater than or equal to 5. They also prove that, in the P_3 convexity, deciding whether the percolation time is greater than or equal to 1, 2, or 3 is polynomial in general graphs and deciding whether the percolation time is greater than or equal to 1, 2, 3, or 4 is polynomial in bipartite graphs, completely solving the question of the complexity of the percolation time in the P_3 convexity.

Theorem 5.6 (Benevides et al. 2015) *The problem of deciding whether the percolation time in the P_3 convexity is greater than or equal to k is* NP*-complete for any integer $k \geq 4$.*

Proof Reduction from the 3-SAT problem (see Appendix B). Given m clauses $C = \{C_1, \dots, C_m\}$ with variables $X = \{x_1, \dots, x_n\}$, we denote the three literals of C_i by $\ell_{i,1}$, $\ell_{i,2}$, and $\ell_{i,3}$. Next, we construct a graph G according to this 3-SAT instance.

For each clause C_i, add to G the subgraph of Fig. 5.2. For each pair of literals $\ell_{i,a}, \ell_{j,b}$ such that one is the negation of the other, add a vertex $y_{(i,a),(j,b)}$ adjacent to $w_{i,a}$ and $w_{j,b}$. Let Y be the set of all vertices created in this way. In addition,

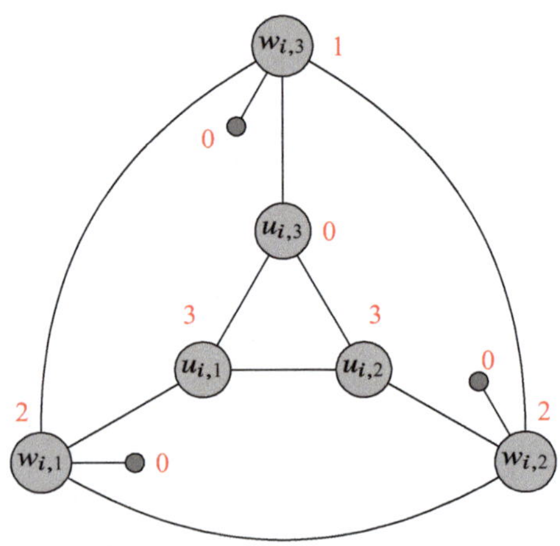

Fig. 5.2 Subgraph added to G for each clause C_i. The red numbers outside the vertices indicate the iteration time of the vertex from the set of vertices with time 0

add a vertex z adjacent to all vertices in Y and a vertex z' of degree 1 whose only neighbor is z. Denote the sets $\{u_{i,1}, u_{i,2}, u_{i,3}\}$ and $\{w_{i,1}, w_{i,2}, w_{i,3}\}$ by U_i and W_i, respectively. Let $U = \cup_{1 \le i \le m} U_i$ and $W = \cup_{1 \le i \le m} W_i$. Let L be the set of vertices of degree 1 in G.

First, consider the case $k = 4$. We show that C is satisfiable if and only if G contains a hull set with percolation time at least 4.

Suppose that C is satisfiable (it has an assignment of the variables that satisfies every clause). Therefore, for every clause C_i, let $k_i \in \{1, 2, 3\}$ such that ℓ_{i,k_i} is true. Let $S' = \{u_{i,k_i} : 1 \le i \le m\}$ and $S = S' \cup L$. Then $I^1(S)$ is obtained from S by adding the vertices $\{w_{i,k_i} : 1 \le i \le m\}$; $I^2(S)$ is obtained from $I^1(S)$ by adding the remaining vertices of W; $I^3(S)$ is obtained from $I^2(S)$ by adding the vertices of Y along with the remaining vertices of U; and $I^4(S)$ is obtained from $I^3(S)$ by adding the vertex z. Therefore, G has a percolation time of at least 4.

Now suppose that $\text{tp}(G) \ge 4$ and let S be any hull set of G with percolation time at least 4. Note that $L \subseteq S$. In addition, for every clause C_i, note that $U_i \cap S \neq \emptyset$ because $|N(u_{i,j}) - U_i| \le 1$ for every i, j. This implies that $W \subseteq I^2(S)$, $U \cup Y \subseteq I^3(S)$ and $z \in I^4(S)$. Finally, if $Y \cap I^2(S) \neq \emptyset$, then $z \in I^3$ and $\text{tp}(S) \le 3$, leading to a contradiction. Therefore, $Y \cap I^2(S) = \emptyset$, which implies that no pair $\{u_{i,a}, u_{i,b}\}$, with $\ell_{i,a}$ being the negation of $\ell_{j,b}$, is in S. Therefore, assigning true to each literal $\ell_{i,j}$ such that $u_{i,j} \in S$ results in an assignment that satisfies C.

Finally, for values $k > 4$, it is sufficient to subdivide the edge zz' into a path P of length $k - 4$, including a new vertex of degree 1, neighbor to each vertex of P. □

Theorem 5.7 (Marcilon and Sampaio 2018b) *In bipartite graphs, the problem of deciding whether the percolation time in the P_3 and P_3^* convexities is greater or equal to k is* NP*-complete for any integer $k \ge 5$.*

Proof The idea is to follow the proof of Theorem 5.6, replacing the subgraph of Fig. 5.2 with that of Fig. 5.3, for each clause C_i. The rest of the construction of the graph G is identical. It is easy to see that the graph G constructed is bipartite. Note also that, similarly to the proof of Theorem 5.6, the vertices inside the square of Fig. 5.3 form a coconvex set, since every vertex has at most one neighbor outside. That is, every hull set must have at least one vertex from each square in G.

In Theorem 5.6, the vertices $w_{i,j}$ are generated at time 1 or 2, which causes the vertex z to be generated at time 4 if the formula is satisfiable and time 3 otherwise. With the substitution for the subgraph of Fig. 5.3, the vertices $w_{i,j}$ will be generated at time 1, 2, or 3, which means that, following the arguments of the proof of Theorem 5.6, the vertex z will be generated at time 5 if the formula is satisfiable and time 4 otherwise. □

With Theorems 5.6 and 5.7, we conclude the following:

Corollary 5.1 *In the P_3 convexity, the percolation time is $(4/3-\varepsilon)$-inapproximable in general graphs and $(5/4 - \varepsilon)$-inapproximable in bipartite graphs for all $\varepsilon > 0$.*

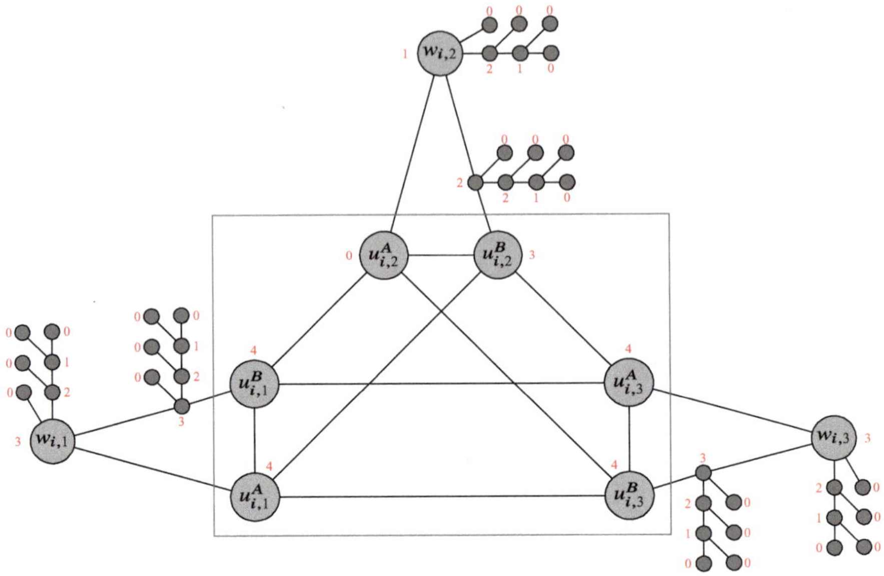

Fig. 5.3 Bipartite subgraph added to G for any clause C_i. Red numbers outside vertices indicate the iteration time of the vertex from the set of vertices with time 0

Proof Gap technique to prove inapproximability: if there is a polynomial algorithm with approximation factor r less than $4/3$, then it can be used to decide between time 3 and 4. The same for $5/4$ in bipartite graphs. □

Exercises

Exercise 5.1 Show that, if G is the graph obtained from the graph with $n-2$ vertices and no edge, adding 2 universal vertices, then $\text{con}_{\text{p3}}(G') = 1$ and $\text{con}_{\text{p3}*}(G') = n-1$.

Exercise 5.2 Complete the proof of Theorem 5.2.

Exercise 5.3 Theorem 5.4 obtains an equation for the rank in the graph G'_u. Obtain the equations of the other convexity parameters.

Exercise 5.4 Let G' be the graph obtained from $n-2$ isolated vertices, adding 2 universal vertices. Prove that $\text{con}_{\text{p3}}(G') = 1$ and $\text{con}_{\text{p3}*}(G') = n-1$.

Exercise 5.5 Given an instance $(U, \mathcal{F}, k)$ of the **Parameterized Hitting Set** problem, where $|U| = n$ and $|\mathcal{F}| = m$, specify the order of the graph G constructed in the proof of Theorem 5.5 in terms of n and m.

Chapter 6
Geodesic Convexity

In this chapter, we present some results about the parameters of Chap. 3 in the geodesic convexity.[1] There is no intention here to present a complete literature review on this convexity, which is very extensive. To the interested reader, the book by Pelayo (2013) is recommended for references up to 2013. However, we note that there is a lot of recent literature published on the subject, especially on computational complexity. We give special focus to these works.

It should be remembered that, as presented in Chap. 2, the geodesic convexity in a graph G is an interval convexity, whose interval function is defined as follows: $I_g(S) = S \cup \{v \in V(G) \mid \exists u, w \in S : v$ belongs to some u, w-geodesic in $G\}$, for all $S \subseteq V(G)$. Remember that u, v-geodesic means u, v-shortest path. That is, the geodesic convexity in G is the family of subsets $\mathcal{C} \subseteq 2^{V(G)}$ such that $S \in \mathcal{C}$ if and only if $I_g(S) = S$. The elements of $\mathcal{C}$ are said to be g-*convex*.

The convex hull of $S \subseteq V(G)$ in the geodesic convexity, denoted by $\text{conv}_g(S)$, can then be obtained by successive applications of the interval function $I_g(\cdot)$. Algorithmically, to obtain $\text{conv}_g(S)$, one can initially add S to a candidate set C and iteratively repeat the following procedure: while there is a vertex $w \in V(G) \setminus C$ that belongs to some u, v-shortest path between two vertices $u, v \in C$, add w to C. Such an algorithm can be executed in $O(|S| \cdot m)$ (Dourado et al. 2009), where m is the number of edges of G. See Exercise 6.1.

Lemma 3.1 of page 20 shows that the *simplicial vertices* of a graph are the extreme vertices in the geodesic convexity. That is, $V(G) \setminus \{v\}$ is g-convex if and only if v is simplicial. In other words, there are no two vertices u and w different from v such that v is in a shortest path between u and w.

---

[1] Some authors, like Pelayo (2013), use the term *geodetic* as adjective of the noun *geodesic*. However, classic books like Van de Vel (1993) and Gruber and Wills (1993) do not use the term *geodetic*, using *geodesic* as a noun and adjective. For example, *geodesic interval* and *geodesic convexity*. We follow the standard of Van de Vel (1993); Gruber and Wills (1993).

J. Araújo et al., *Introduction to Graph Convexity*, SBMAC Collection on Applied and Computational Mathematics, https://doi.org/10.1007/978-3-031-84128-6_6

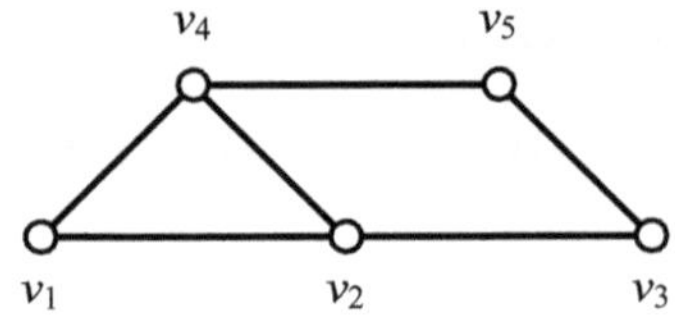

Fig. 6.1 Example for the geodesic convexity

A subset S is said to be *coconvex* in the geodesic convexity of a graph G, if $V(G) \setminus S$ is g-convex. Clearly, the unitary sets that contain exactly one simplicial vertex of a graph are coconvex in the geodesic convexity. This fact is of great importance for some proofs in the rest of the chapter.

In Fig. 6.1, the vertex v_1 is simplicial and, therefore, the subset $\{v_2, v_3, v_4, v_5\}$ is g-convex. Furthermore, note that $\{v_1, v_2, v_3\}$ is also g-convex, since there is no shortest path between two of these vertices that has v_4 or v_5 as an internal vertex. Consequently, the subset $\{v_4, v_5\}$ is also coconvex, as is $\{v_1\}$. Lastly, also note that the subset $\{v_4, v_5\}$ itself is g-convex.

Another relevant fact previously mentioned is that there is a characterization of Ptolemaic graphs presented in Theorem 4.2. These are exactly the graphs in which the geodesic convexity is a convex geometry, as seen in Sect. 4.3.

6.1 Interval and Hull Numbers

As in Sects. 4.2 and 4.3, an interval set $S \subseteq V(G)$ in the geodesic convexity, also called *geodesic set*, is such that $\mathrm{I_g}(S) = V(G)$. The interval number in the geodesic convexity of G is the cardinality of a smallest geodesic set of G. Such a parameter is also known as the *geodesic number* and denoted by $\mathrm{in_g}(G)$ (Harary et al. 1993). A hull set $S \subseteq V(G)$ in the geodesic convexity of G satisfies $\mathrm{conv_g}(S) = V(G)$. The cardinality of a smallest hull set in the geodesic convexity of G, denoted by $\mathrm{hn_g}(G)$, is the hull number of G (Everett and Seidman 1985).

As an example, the reader can review Sects. 4.2 and 4.3, page 38. If G is the graph represented in Fig. 6.1, note that v_1 must belong to every interval set and to every interval set in the geodesic convexity of this graph, as it is simplicial. Moreover, there is no hull set with two vertices and, consequently, no interval set with two vertices in the geodesic convexity of G. Trivially, $\{v_1, v_3, v_5\}$ is an interval set and, therefore, it is also a hull set, so $\mathrm{in_g}(G) = \mathrm{hn_g}(G) = 3$. Next, we present some results about these two parameters, which are the most studied in the literature.

Bounds and Characterizations

The *diameter* of a graph G is the size of a maximum shortest path between two vertices of G, denoted by $\mathrm{diam}(G)$. It follows that:

Theorem 6.1 (Chartrand et al. 2002c) *If G is nontrivial, then*

$$2 \leq \text{hn}_g(G) \leq \text{in}_g(G) \leq n(G) - \text{diam}(G) + 1.$$

Proof See Exercise 6.2. □

As previously stated, due to Lemma 3.1, every simplicial vertex in a graph G must belong to every interval set and hull set of G in the geodesic convexity. More than that, if $v \in V(G)$ is not simplicial, then $V(G) \setminus \{v\}$ is an interval set and, therefore, a hull set. Therefore, $\text{hn}_g(G) = \text{in}_g(G) = n(G)$ if and only if G is complete. In the case of the hull number, such fact is observed by Everett and Seidman (1985).

A natural question is: in which graph classes these parameters are equal? As stated in Sect. 4.3, such equality occurs in trees and cycles.

Theorem 6.2 (Chartrand et al. 2002c) *For all n, k and d integers such that $n - d - k + 1 \geq 0$, $2 \leq k \leq n$, and $2 \leq d \leq n$, there exists a graph G such that $\text{hn}_g(G) = \text{in}_g(G) = k$, $\text{diam}(G) = d$, and $n(G) = n$.*

Proof Let G_k be the graph represented in Fig. 6.2, obtained from a path $(u_0, \ldots, u_d)$ by adding $k - 2$ vertices $v_1, \ldots, v_{k-2}$ of degree one, whose only neighbor is u_1, and $n - d - k + 1$ vertices $w_1, \ldots, w_{n-d-k+1}$, whose neighborhood is exactly $\{u_0, u_2\}$. Note that $n(G) = n$, $\text{diam}(G) = d$, and the vertices $v_1, \ldots, v_{k-2}, u_d$ are simplicial. Therefore, they are trivial coconvex sets, but $S = \{v_1, \ldots, v_{k-2}, u_d\}$ is neither an interval set nor a hull set. Then, $\text{hn}_g(G_k) \geq k$ and $\text{in}_g(G_k) \geq k$, with these inequalities being satisfied in equality since if we add u_0 to the aforementioned subset S, we obtain an interval set in the geodesic convexity. □

Chartrand et al. (2000) show a family of graphs G such that $\text{hn}_g(G) = \alpha$ and $\text{in}_g(G) = \beta$ for all $2 \leq \alpha \leq \beta \leq n(G)$ integers. See Exercise 6.3.

There are also in the literature several bounds for the hull number in the geodesic convexity. Some examples of these bounds can be found in the works of Everett and Seidman (1985) and Dourado et al. (2010a). Next, we detail a more recent proof, which improves a bound of Everett and Seidman (1985) and answers a question left open in the same work.

The proof of such bounds uses in general the same idea. A hull set S of a given graph G is constructed iteratively. Given some bounds, two vertices at maximum distance are added to S initially, which implies that, in the first application of the

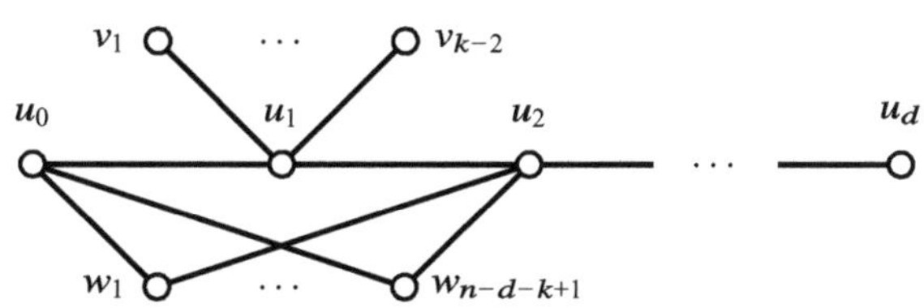

Fig. 6.2 Graph G_k

interval function on S, at least the vertices of a maximum shortest path belong to the hull of S. Such bounds, consequently, are written in terms of $\text{diam}(G)$. Other bounds initially add to S the set of simplicial vertices, if they exist and, if not, just an arbitrary vertex. Then, the following process is repeated: while the hull of S is distinct from $V(G)$, a vertex is sought in $V(G) \setminus \text{conv}_g(S)$ to be added to S so that it is guaranteed that more vertices in $V(G) \setminus \text{conv}_g(S)$ belong to the hull of the new set S. We present one of these bounds below.

Theorem 6.3 (Araújo et al. 2013) *Let G be a connected graph with n vertices and s simplicial vertices. Then,*

$$\text{hn}_g(G) \leq \max\{1, s\} + \left\lceil \frac{3(n - \max\{1, s\})}{5} \right\rceil.$$

Furthermore, this bound is tight.

Proof We describe below an algorithm that iteratively constructs a hull set S of G whose cardinality satisfies such a bound.

If G has simplicial vertices, define $S_0 \subseteq V(G)$ as the subset of simplicial vertices of G. Otherwise, define $S_0 \subseteq V(G)$ as a unitary set by arbitrarily choosing a vertex of G to belong to S_0. Note that this choice corresponds to the terms $\max\{1, s\}$ in the bound above. Define $H_0 = \text{conv}_g(S_0)$. Consider $i = 0$ initially. If $H_0 = V(G)$, there is nothing more to add and we return $S = S_0$.

Let us describe how to add a subset of chosen vertices X_i to S_i at step $i + 1$ such that $S_{i+1} = S_i \cup X_i$ and we take $H_{i+1} = \text{conv}_g(S_{i+1})$. We denote by $\overline{H_i} = V(G) \setminus H_i$. Moreover, the choice of X_i should imply the addition of new vertices in H_{i+1} in a ratio of 3 to 5, i.e., for every 5 vertices of H_{i+1}, at most 3 belong to S_{i+1}, so that at the end of the process the desired bound is obtained.

While $\overline{H_i} \neq \emptyset$, if there is a vertex $u \in \overline{H_i}$ whose shortest path to some vertex in H_i has length at least 2, then take $X_i = \{u\}$ and the sets S_{i+1} and H_{i+1} are defined accordingly. Note that there is at least one vertex in $\overline{H_i}$ not included in S_{i+1}, which belongs to H_{i+1}, that one belonging to a shortest path from u to some vertex of H_i. See Fig. 6.3a. Therefore, one vertex was selected for S_{i+1} and at least one more is included in H_{i+1}, so the ratio of 3 to 5 is respected.

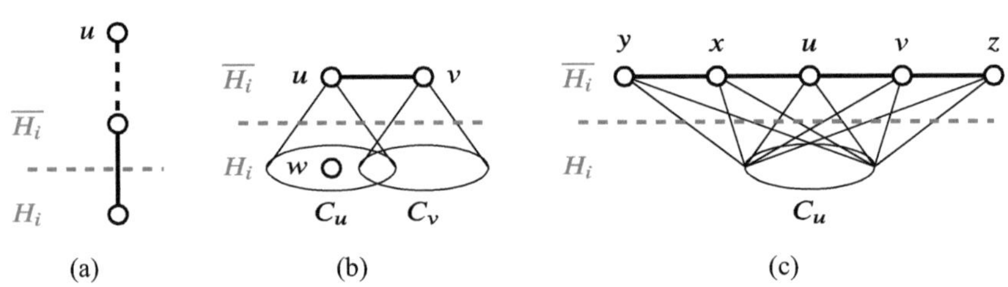

Fig. 6.3 Case analysis for Theorem 6.3

If there is no such vertex, then every vertex in $\overline{H_i}$ has at least one neighbor in H_i. It should be noted that, for each $u \in \overline{H_i}$, we have that $N_G(u) \cap H_i$ is necessarily a clique, since H_i is g-convex and $u \notin H_i = \text{conv}_g(S_i)$. Moreover, as the simplicial vertices of G belong to S_0, if they exist, then u cannot be simplicial. Consequently, u has a neighbor $v \in \overline{H_i}$.

Denote by $C_u = N(u) \cap H_i$ the neighborhood of u in H_i which, as said, is a clique. By the same argument, v also has neighbors in H_i and such neighborhood corresponds to a clique C_v. If there is a vertex $w \in C_u \setminus C_v$, note that we can in this case choose $X_i = \{v\}$, since u belongs to a shortest v, w-path and we again add a vertex to S_{i+1} and one more vertex is included in H_{i+1}. See Fig. 6.3b. The analogous argument occurs when there is a vertex $w \in C_v \setminus C_u$. Therefore, let us assume that $C_v = C_u$ and u must have a neighbor $x \in \overline{H_i}$ such that $xv \notin E(G)$. Analogously, we can argue that x must have neighbors in H_i, that $N(x) \cap H_i = C_x$ is a clique and $C_x = C_u = C_v$. Without loss of generality, we can assume that u is a vertex of minimum degree in $\overline{H_i}$. Consequently, x and v must have neighbors y and z, respectively, such that $y, z \notin N(u)$. See Fig. 6.3c. In this case, we add to S_{i+1} not just one, but three vertices at once, taking $X_i = \{u, y, z\}$ and noting that x and v belong to H_{i+1}. If $H_{i+1} = V(G)$, we return $S = S_{i+1}$. If not, we increment the value of i by one and restart the analysis.

For a tight example, take the union of several C_5 and add a universal vertex to them. It is easy to see that 3 out of every 5 vertices of each C_5 belong to every hull set of such a graph. This example is due to Everett and Seidman (1985). □

Next, we present computational complexity results for both parameters $\text{in}_g(G)$ and $\text{hn}_g(G)$ even in relatively simple graph classes, as well as we present polynomial time algorithms in other graph classes for both parameters. Given a graph G and a positive integer k, it is NP-complete to decide whether $\text{in}_g(G) \leq k$ and whether $\text{hn}_g(G) \leq k$. Below, we present a brief summary of complexity results for these parameters and examples of proofs.

Hardness of $\text{in}_g(G)$

The computational hardness of determining the geodesic interval number was proved in the first paper on the subject (Harary et al. 1993). Dourado et al. (2010b) show that it is still NP-hard to determine $\text{in}_g(G)$, even if G is chordal or chordal bipartite. It is also NP-hard to determine $\text{in}_g(G)$, when G is P_5-free (Dourado et al. 2016c). A similar result was obtained for planar graphs and line graphs by Chakraborty et al. (2020b), while Chakraborty et al. (2020a) show the hardness for partial grid subcubic graphs and interval graphs, improving previous results for subcubic graphs (Bueno et al. 2018) and for chordal graphs (Dourado et al. 2010b). Kellerhals and Koana (2022) study the parameterized complexity of this parameter and show that deciding whether $\text{in}_g(G) \leq k$ is W[1]-hard when parameterized by k, the size of a minimum vertex cover of G, and the pathwidth of G combined.

We show below one of the simplest reductions found in the literature for this parameter. Given a graph G, a subset $S \subseteq V(G)$ is *dominating* if $N[v] \cap S \neq \emptyset$ for every vertex v of G, i.e., every vertex is in S or has a neighbor in S. The cardinality of a smallest dominating set of G is denoted by $\gamma(G)$. The problem **Dominating Set** below is one of the most classic NP-complete problems (Garey and Johnson 1979).

Dominating Set

Instance: Graph G and an integer $k \geq 0$.
Question: $\gamma(G) \leq k$?

We reduce this problem to that of **Geodesic Interval Number**.

Geodesic Interval Number

Instance: Graph G and an integer $k \geq 0$.
Question: $\mathrm{in_g}(G) \leq k$?

Theorem 6.4 (Dourado et al. 2010b) *Given a graph G and a positive integer k, **Geodesic Interval Number** is NP-complete, even if G is chordal.*

Proof We left as Exercise 6.4 the proof that **Geodesic Interval Number** is in NP. Given an instance (G, k) of **Dominating Set**, we construct in linear time an equivalent instance (G', k') of **Geodesic Interval Number**, where[2] $k' = k + n(G)$.

The graph G' is obtained from G, i.e., first, G is copied to be a subgraph of G'. For each vertex $v \in V(G)$, two new vertices $x_v, y_v \in V(G')$ and the edges $y_v x_v, x_v v \in E(G')$ are added to G'. Finally, a vertex $z \in V(G')$ is added to G' that is adjacent in G' to all vertices v and x_v, for all $v \in V(G)$. Note that $n(G') = 3n(G) + 1$ and, therefore, G' can be constructed in linear time. It is up to the reader to verify that G' is chordal; see Exercise 6.5. See Fig. 6.4 for a small example of construction on a graph $G = P_3$ with vertices a, b, and c. Define $X = \{x_v \mid v \in V(G)\}$ and $Y = \{y_v \mid v \in V(G)\}$.

We must prove that $\gamma(G) \leq k$ if and only if $\mathrm{in_g}(G') \leq k + n(G)$. Suppose first that S is a dominating set of G such that $|S| \leq k$. Define $S' = S \cup Y$. Therefore, $|S'| \leq k + n(G)$. We will prove that S' is a geodesic interval set of G. Note that

[2] As k' is not exclusively a function of k, note that such a reduction cannot be used to argue that **Geodesic Interval Number** is W[2]-hard, despite this fact being true for **Dominating Set**.

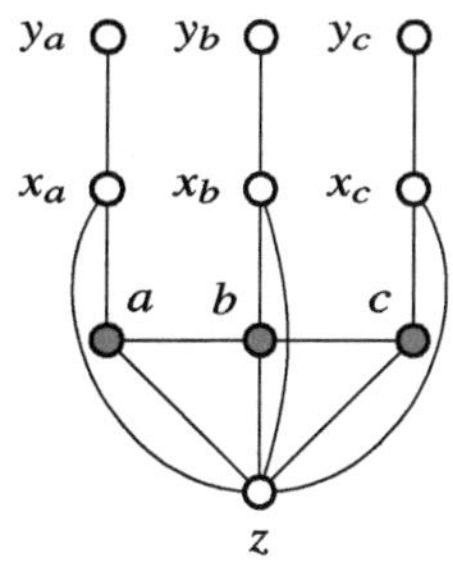

Fig. 6.4 Construction of G' from $G = P_3$. Vertices of G are filled in gray

$X \cup \{z\} \subseteq \mathrm{I_g}(Y) \subseteq \mathrm{I_g}(S')$, since there is a y_v, y_u-shortest path (y_v, x_v, z, x_u, y_u) in G' for all $v, u \in V(G)$ with $v \neq u$, also assuming, without loss of generality, that G is connected and nontrivial. As S is dominating, also note that each vertex of $v \in V(G) \setminus S$ has some neighbor $u \in S$. Therefore, in G', $v \in \mathrm{I_g}(S')$, since $u \in S \subseteq S'$, $y_v \in S'$, and (u, v, x_v, y_v) is a shortest path of G'.

Assume now that S' is a geodesic interval set of G' such that $|S'| \leq k+n(G)$. Let $S = S' \cap V(G)$. We know that $Y \subseteq S'$; see Lemma 3.1. Therefore, $|S| \leq k$. We will prove that S is a dominating set of G. Let $u \in V(G) \setminus S$. As S' is a geodesic interval set of G', there exist vertices $v, w \in S'$ such that u belongs to some v, w-shortest path of G'. We will prove that $\{v, w\} \cap N_G(u) \neq \emptyset$ and, therefore, the result follows.

It is not possible that $\{v, w\} \subseteq Y$, since there is only one y_a, y_b-shortest path in G', which contains x_a, z, and x_b, i.e., it does not contain any vertex of $V(G)$ in G'. If $\{v, w\} \subseteq X \cup V(G)$, note that both v and w are neighbors of z. The existence of a v, w-path that contains u implies that $\mathrm{dist}_{G'}(v, w) = 2$, $v \neq z$, and $w \neq z$. Therefore, $\{v, w\} \subseteq N_{G'}(u)$ and consequently $N_G(u) \cap \{v, w\} \neq \emptyset$, since $v \neq z$, $w \neq z$, and u has only one other neighbor in $V(G') \setminus V(G) = \{x_u\}$. Finally, suppose that $v \in Y$ and $w \in X \cup V(G)$. As any two vertices of $X \cup V(G)$ are at distance at most 2 in G' and $v = y_a$ has a single neighbor x_a in G', for some $a \in V(G)$, we have that $\mathrm{dist}_{G'}(v, w) = 3$ and the path $(v = y_a, x_a, u, w)$ is a v, w-shortest path of G'. In this case, the only possibility is $u = a$ and $w \in N_G(a)$. □

Although the authors do not explicitly mention it in their paper, the previous proof shows that Theorem 6.4 holds even for graphs with diameter at most 4.

Tractable Cases for $\mathbf{in_g}(G)$

Despite the computational hardness presented above, the parameter $\mathrm{in_g}(G)$ can be calculated in polynomial time for various graph classes: cographs (Dourado et al. 2010b), split graphs (Dourado et al. 2010b), Ptolemaic graphs (Farber and Jamison 1986), block cactus graphs (Ekim et al. 2012), periplanar graphs (Mezzini 2018), and proper interval graphs (Ekim et al. 2012). In addition, the value of $\mathrm{in_g}(G)$ can be solved in FPT time when parameterized by the treedepth of G, by the feedback edge set number of G, and by the modular-width of G (Kellerhals and Koana 2022).

Chakraborty et al. (2020a) further show that deciding whether $\text{in}_g(G) \leq k$ can be solved in FPT time when parameterized by the treewidth of G. Such an algorithm implies a linear algorithm for trees.

Next, we present an example of a polynomial algorithm for determining the geodesic interval number. The class of *cographs* is defined recursively as follows:

1. A trivial graph is a cograph.
2. If G is a cograph, then $\overline{G}$ is also a cograph.
3. If G_1 and G_2 are cographs, then the disjoint union $G = G_1 \cup G_2$ is a cograph.

These graphs are widely studied in the literature. There are various characterizations of them, perhaps the most well known being that they correspond to graphs that do not have paths with 4 vertices as an induced subgraph (Corneil et al. 1981). By the recursive definition of the class, for a cograph G, one can obtain in polynomial time a rooted tree (T, r) that represents the construction of G, whose leaves are the vertices of G and each internal node $v \in V(T)$ represents one of the operations (complement or disjoint union) applied to the subgraphs represented by the children of v. The graph G is then the cograph represented by the root r of T. This is the modular decomposition tree of G and it can be obtained in linear time (McConnell and Spinrad 1999).

Theorem 6.5 (Dourado et al. 2010b) *If G is a connected cograph and $\overline{G}$ has k nontrivial connected components, whose subgraphs induced by the vertices of these components in G are $G_1, \ldots, G_k$, then:*

1. *If* $k = 0$, *then* $\text{in}_g(G) = n(G)$.
2. *If* $k = 1$, *then* $\text{in}_g(G) = \text{in}_g(G_1)$.
3. *If* $k \geq 2$, *then* $\text{in}_g(G) = \min\{4, \min\{\text{in}_g(G_i) \mid i \in \{1, \ldots, k\}\}\}$.

Proof To prove Item 1, note that under this hypothesis G must be a complete graph. Therefore, $\text{hn}_g(G) = n(G)$, by Lemma 3.1.

To prove Item 2, observe that G_1 is neither a complete subgraph nor trivial, since G_1 corresponds to a nontrivial connected component in $\overline{G}$. Furthermore, as G is a cograph, G does not have induced subgraphs isomorphic to a P_4. The same occurs for G_1. Therefore, any two vertices in G_1 are at a maximum distance of 2, since every shortest path is induced. Therefore, a minimum geodesic interval set S_1 of G_1 must contain two nonadjacent vertices, since G_1 is not complete, and $\text{I}_g(S_1)$ is obtained from paths of length 2. Therefore, S_1 is also a geodesic interval set of G. Therefore, $\text{in}_g(G) \leq \text{in}_g(G_1)$. On the other hand, observe that any minimum geodesic interval set of G cannot contain a vertex $v \in V(G) \setminus V(G_1)$, since v is universal in G (see Exercise 6.6). Therefore, $\text{in}_g(G) = \text{in}_g(G_1)$.

Finally, to prove Item 3, note that each G_i is a subgraph of G that is neither complete nor trivial and, as the vertices of these subgraphs form components in $\overline{G}$, there are all edges between any two vertices of distinct subgraphs G_i and G_j for $1 \leq i \neq j \leq k$.

Let $S = \{v_1, v_2, u_1, u_2\}$, for any two nonadjacent vertices $v_1, v_2 \in V(G_1)$ and any two nonadjacent vertices $u_1, u_2 \in V(G_2)$. Note that every vertex $w \in V(G) \setminus$

$V(G_1)$ satisfies $w \in \mathrm{I_g}(S)$, since w is adjacent to v_1 and v_2. Similarly, $V(G_1) \in \mathrm{I_g}(S)$, since every vertex of $V(G_1)$ is adjacent to u_1 and u_2. Therefore, $\mathrm{in_g}(G) \leq 4$.

If there is a G_i such that $\mathrm{in_g}(G_i) < 4$, as each G_i is neither trivial nor complete, note that at least two nonadjacent vertices of G_i belong to every minimum interval set S_i in the geodesic convexity of G_i. Therefore, S_i is also a minimum geodesic interval set of G. Conversely, we will prove that if G has a geodesic minimum interval set S of cardinality less than 4, such a set corresponds to a minimum hull set for G_i for some $i \in \{1, \ldots, k\}$. If $S \subseteq V(G_i)$, there is nothing to prove. Suppose the contrary. If $|S| = 2$, then the two elements of S are adjacent. S could only be an interval set of G, if $V(G) = S$, but G is not complete. If $|S| = 3$, again as G is not complete, the only possibility is S having two elements in a subgraph G_i and one in G_j for $1 \leq i \neq j \leq k$. Let $v_1, v_2 \in S \cap V(G_i)$ and $u \in S \cap V(G_j)$. As $u \in \mathrm{I_g}(\{v_1, v_2\})$ and $uv_1, uv_2 \in E(G)$, then $S \setminus \{u\}$ is a minimum geodesic interval set of G, contradicting the minimality of S. Therefore, this last case does not occur. □

Hardness of $\mathbf{hn_g}(G)$

Regarding the geodesic hull number, the first result in the literature that deals with the computational hardness of determining this parameter was presented by Dourado et al. (2009).

Such reduction was modified to show that it is NP-complete to decide whether $\mathrm{hn_g}(G) \leq k$ even if G is restricted to be bipartite, but it is still a considerably long proof (Araújo et al. 2013). There is also in the literature another proof of NP-hardness for determining $\mathrm{hn_g}(G)$ when G belongs to the class of partial cubes, a subclass of bipartites (Albenque and Knauer 2016). Such reduction strongly relies on properties inherent to this graph class and it is also not simple. It was also shown that determining $\mathrm{hn_g}(G)$ is NP-hard for the class of P_9-free graphs (Dourado et al. 2016c).

The difficulty in presenting these results lies in the fact that it is necessary to check all the shortest paths between the pairs of vertices of the instance constructed by the reduction.

The simplest way to prove that deciding whether $\mathrm{hn_g}(G) \leq k$, for any graph G, is a computationally hard problem, which we found in the literature, uses a similar result for $\mathrm{hn_p}(G)$ when G is triangle-free and the observation already presented in Lemmas 2.3 and 2.4 on Page 15, whose essential argument we present below.

Lemma 6.1 *Let G be a triangle-free graph that is not complete and let G' be the graph obtained from G by adding a universal vertex v'. Then S is a minimum hull set in the P_3 convexity of G, if and only if, S is a minimum hull set in the geodesic convexity of G.*

Proof First, it should be noted that, as G is free of triangles, each vertex $u \in \mathrm{I}^k_{\mathrm{p3}}(S) \setminus \mathrm{I}^{k-1}_{\mathrm{p3}}(S)$ is added to $\mathrm{conv}_{\mathrm{p3}}(S)$ in the kth iteration in G due to the existence of an *induced* P_3 path in G such that u is an internal vertex of an x, y-path P_3, for vertices $x, y \in \mathrm{I}^{k-1}_{\mathrm{p3}}(S)$. As G' has a universal vertex v', any two vertices $x, y \in V(G)$ are at distance at most 2 in G'. Therefore, every shortest path between two vertices $x, y \in V(G)$ in G' is an induced path of G of length at most 2. Therefore, S is also a hull set in the geodesic convexity of G', since, as G' is not complete, v' belongs to the hull of any two nonadjacent vertices of S in G'.

For the converse, let S' be a minimum hull set in the geodesic convexity of G'. As G is not complete, G' is also not. As v' is universal of G' and S' is a *minimum* hull set in the geodesic convexity of G', note that v' does not belong to S' (see Exercise 6.6). Therefore, $S' \subseteq V(G)$. We claim that S' is a hull set in the P_3 convexity of G. As S' is a hull set in the geodesic convexity of G' and G' has diameter 2, it follows that each shortest path in G' has length at most 2 and, therefore, any $u \in \mathrm{I}^k_{\mathrm{g}}(S') \setminus \mathrm{I}^{k-1}_{\mathrm{g}}(S')$ is added to $\mathrm{conv}_{\mathrm{g}}(S')$ in the kth iteration in G' due to the existence of an *induced* P_3 path in G such that u is an internal vertex of an x, y-path P_3, for vertices $x, y \in \mathrm{I}^{k-1}_{\mathrm{g}}(S')$. Therefore, the result follows. □

With Lemma 6.1, we can argue the computational hardness of the following problem.

Parametrized Geodesic Hull Number

Instance:	A graph G and an integer $k \geq 0$.
Parameter:	k.
Question:	$\mathrm{hn}_{\mathrm{g}}(G) \leq k$?

Corollary 6.1 (Kanté et al. 2019) *The problem* ***Parametrized Geodesic Hull Number*** *is* W[2]*-hard, even when G has diameter 2.*

Proof It is a direct consequence of Theorem 5.5, presented on Page 60, and of Lemma 6.1. □

Kanté et al. (2019) also show that deciding whether $\mathrm{hn}_{\mathrm{g}}(G) \leq k$ is W[1]-hard when parameterized by k and the treewidth combined, and it is XP when parameterized only by the treewidth of G.

Tractable Cases for $\mathbf{hn_g(G)}$

The geodesic hull number, despite being intractable even for subclasses of bipartite graphs such as partial cubes, can be determined in polynomial time for some graph classes. Dourado et al. (2009) show polynomial algorithms to determine $\mathrm{hn_g}(G)$ when G is a unit interval graph, when G is split or when G is a cograph. Araújo et al. (2013) generalize the result of cographs to the class of $(q, q-4)$-graphs, which are the graphs such that each subset of at most q vertices induces at most $q-4$ P_4's. They also obtain polynomial algorithms for complements of bipartite graphs and for cacti. Kanté and Nourine (2013) present a polynomial algorithm for distance–hereditary graphs. Dourado et al. (2016c) show polynomial algorithms to determine $\mathrm{hn_g}(G)$ when G is paw-free and P_5-free, when every 6 vertices of G induce at most one P_5 and when G is P_k-free and the girth of G is at least $k-1$ for every positive integer k. Coelho et al. (2022) recently presented a polynomial algorithm to calculate $\mathrm{hn_g}(G)$ when G is a complementary prism.

We leave to the reader the proof of the following result, which implies a polynomial algorithm for determining the geodesic hull number of a cograph. The proof is quite similar to the proof of Theorem 6.5.

Theorem 6.6 (Dourado et al. 2009) *If G is a connected cograph and $\overline{G}$ has k nontrivial connected components, then:*

1. *If $k = 0$, then* $\mathrm{hn_g}(G) = n(G)$.
2. *If $k = 1$, then* $\mathrm{hn_g}(G) = \mathrm{hn}(G_1)$, *where $G_1 \subseteq G$ is the subgraph induced by the vertices of the only nontrivial component of G.*
3. *If $k \geq 2$, then* $\mathrm{hn_g}(G) = 2$.

Proof See Exercise 6.7. □

6.2 Convexity Number

The *geodesic convexity number* $\mathrm{con_g}(G)$, introduced by Chartrand et al. (2002b), is the size of a largest proper g-convex set of G.

There are several computational complexity results for the convexity number. For example, in the geodesic and P_3 convexities, besides the determination of $\mathrm{con_g}(G)$ being an NP-hard problem, it is highly inapproximable, since there is no polynomial algorithm with approximation factor $n^{1-\varepsilon}$ for any $\varepsilon > 0$, unless P = NP (Coelho et al. 2015).

Let us consider the problem of determining the geodesic convexity number in its decision version.

Geodesic Convexity Number

Instance: Graph G and an integer $k \geq 0$.
Question: $\mathrm{con_g}(G) \geq k$?

Let us initially show that we can solve the above problem in polynomial time when k is a constant. By definition, $\text{cong}_g(G) < k$ if and only if the geodesic convex hull of any set with exactly k vertices contains all the vertices of G. Since k is constant, we can generate and test all these sets in polynomial time. Therefore, for constant k, the **Geodesic Convexity Number** is in P.

Now let us show that the **Convexity Number** problem is NP-complete even when the input graph G is bipartite.

Theorem 6.7 (Dourado et al. 2012) *The* ***Geodesic Convexity Number*** *restricted to bipartite graphs is* NP-*complete.*

Proof Since the geodesic convex hull of a set can be determined in polynomial time, the **Geodesic Convexity Number** is in NP. To prove NP-completeness, we reduce an instance (H, k) of the known NP-complete problem **Clique** (see Garey and Johnson 1979, p. 194) to an instance (G, k') of the **Geodesic Convexity Number** such that the graph H has a clique of order at least k if and only if (i) $\text{cong}_g(G) \geq k'$; (ii) the encoding length of (G, k') is polynomially bounded in terms of the encoding length of (H, k) and (iii) G is bipartite.

Let (H, k) be an instance of **Clique**. Clearly, we can assume that H is connected and $k \geq 3$. We construct G in the following way. For each vertex u of H, we create four vertices w_u, x_u, y_u, and z_u in G and add three edges $x_u z_u$, $y_u z_u$, and $w_u z_u$ as shown in Fig. 6.5a. For each edge uv of H, we create a set $V_{uv} = \{a_{uv}, b_{uv}, c_{uv}, d_{uv}, e_{uv}\} \cup I_{uv}$ of $n+5$ additional vertices in G, where n denotes the order of H and I_{uv} denotes an independent set of n vertices, and add edges so that z_u, w_u, z_v, and w_v, together with the vertices in V_{uv}, induce the graph G_{uv} as shown

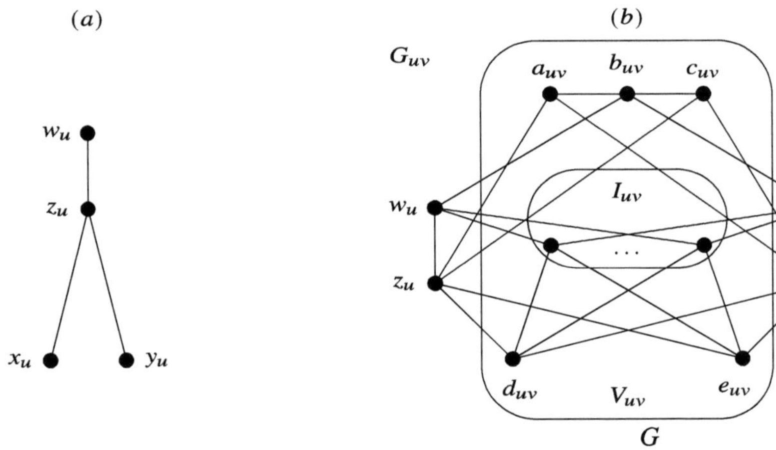

Fig. 6.5 *Gadgets* for the construction of G

in Fig. 6.5b, where all vertices in I_{uv} have exactly the same four neighbors. In other words,

$$\begin{aligned}
V(G_{uv}) &= \{w_u, z_u, w_v, z_v \mid uv \in E(H)\} \cup \{a \in V_{uv} \mid uv \in E(H)\};\\
E(G_{uv}) &= \{z_u w_u, z_v w_v\}\\
&\quad \cup \{a_{uv} b_{uv}, b_{uv} c_{uv} \mid uv \in E(H)\}\\
&\quad \cup \{z_u a_{uv}, z_v a_{uv}, z_u c_{uv}, z_v c_{uv} \mid uv \in E(H)\}\\
&\quad \cup \{z_u d_{uv}, z_v d_{uv}, z_u e_{uv}, z_v e_{uv} \mid uv \in E(H)\}\\
&\quad \cup \{w_u b_{uv}, w_v b_{uv} \mid uv \in E(H)\}\\
&\quad \cup \{a w_u, a w_v, a d_{uv}, a e_{uv} \mid a \in I_{uv}, uv \in E(H)\}.
\end{aligned}$$

To complete the construction, we create two vertices x and y in G and add edges xx_u and yy_u for all vertices u of H. Note that G is bipartite with bipartition (V_1, V_2), where:

$V_1 = \{x_u, y_u, w_u \mid u \in V(H)\} \cup \{a_{uv}, c_{uv}, d_{uv}, e_{uv} \mid uv \in E(H)\}$;

$V_2 = \{z_u \mid u \in V(H)\} \cup \{b_{uv} \mid uv \in E(H)\} \cup \{x, y\} \cup \{a \in I_{uv} \mid uv \in E(H)\}$.

Figure 6.6 illustrates the complete construction of G for the case where H is a path P_3 with three vertices a, b, and c.

Let $k' = 3k + (n + 5)\binom{k}{2} + 1$. Clearly, the length of the encoding of (G, k') is polynomially bounded in terms of the length of the encoding of (H, k). It remains to prove that H has a clique of order at least k if and only if the geodesic convexity number of G is at least k'.

First, we assume that H has a clique C of order at least k and we construct a set S as follows. For every two vertices u and v in C, we add all vertices of G_{uv} to S.

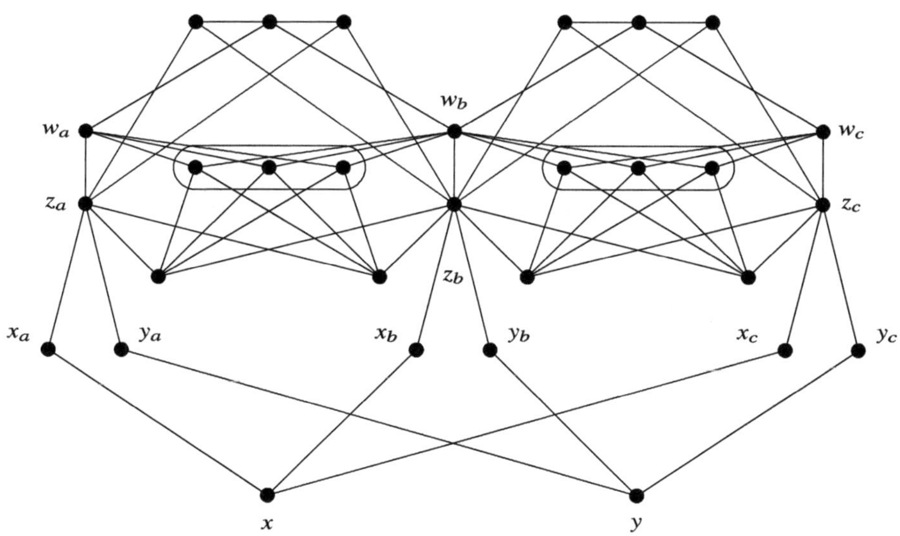

Fig. 6.6 The graph G constructed from $H = P_3$

For each vertex u in C, we add the vertex x_u to S. Finally, we add x to S. It is easy to check that S is a g-convex set, with at least k' vertices, which does not contain y, i.e., $\text{cong}_g(G) \geq k'$.

Next, we assume that G has a g-convex set S of order at least k' that does not contain all vertices of G. We define the following sets:

- $V_x = \{x_u \mid u \in V(H)\}$.
- $V_y = \{y_u \mid u \in V(H)\}$.
- $V_z = \{z_u \mid u \in V(H)\}$.
- $V_w = \{w_u \mid u \in V(H)\}$.

Since $\text{conv}_g(\{x, y\})$ contains all vertices of G, at most one between x and y belongs to S. If S contains more than n vertices of $V_x \cup V_y$, then there exist distinct vertices u and v in H such that x_u and y_v belong to S. As $x, y \in I_g(\{x_u, y_v\})$, we get $x, y \in S$, which is a contradiction. Therefore, S contains at most n vertices of $V_x \cup V_y$.

Claim A: If S contains three vertices of G_{uv} for some edge uv of H, then S contains all vertices of G_{uv}.

Claim A is immediate and does not require proof.

Claim B: S contains at least two vertices of V_z.

Proof of Claim B: By contradiction, let us assume that S contains at most one vertex of V_z. Using Claim A, it follows easily that there is no edge uv of H such that:

- Either $|S \cap \{z_u, w_u\}| + |S \cap V_{uv}| \geq 3$
- Or $|S \cap \{z_u, w_u\}| \geq 1$ and $|S \cap \{z_v, w_v\}| \geq 1$

Similarly, there are no three vertices u, v, and w of H such that uv and vw are edges of H and

$|S \cap \{z_v, w_v\}| = 0$, $|S \cap V_{uv}| \geq 1$ and $|S \cap V_{vw}| \geq 1$.

These observations imply that the set of vertices u of H for which S intersects $\{z_u, w_u\}$ forms an independent set. By hypothesis, there is at most one vertex u of H for which S contains both vertices z_u and w_u. Moreover, if uv and $u'v'$ are two edges of H such that u and u' are distinct and S intersects $\{z_u, w_u\}$, $\{z_{u'}, w_{u'}\}$, V_{uv}, and $V_{u'v'}$, then v and v' are distinct. Finally, if S contains two vertices of V_{uv} for some edge uv of H, then S does not contain any vertex of $\{z_u, w_u, z_v, w_v\}$ or of $V_{uv'}$ for an edge uv' of H different from uv.

These observations easily imply that S contains at most $n + 1$ vertices of

$$V_z \cup V_w \cup \bigcup_{uv \in E(H)} V_{uv}.$$

Together with the previous observations to Claim A, we then conclude that $|S| \leq 2n + 2$. As $k \geq 3$, this is a contradiction. □

Let now $C = \{u \in V(H) \mid z_u \in S\}$. By Claim B, the set C contains at least two elements. If S contains two vertices z_u and z_v such that u and v are not adjacent in H, then the distance of z_u and z_v in G is 4. Therefore, x and y belong to S, which is a contradiction. Therefore, C is a clique of H.

Now assume by contradiction that $|C| = t < k$. Let S' be the union of the vertex sets of the graphs G_{uv} for all pairs of distinct vertices u and v in C. Note that S' contains exactly $2t + (n+5)\binom{t}{2}$ vertices. As S is convex, S' is a subset of S.

Claim C: $S \setminus S'$ does not contain vertices of $V_w \cup \bigcup_{uv \in E(H)} V_{uv}$.

Proof of Claim C By contradiction, we assume that S contains a vertex a from this set.

First, let us assume that $a = w_u$ for some vertex u of H. By the definition of S', $u \notin C$. Let v be some vertex in C. Note that $\mathrm{I_g}(\{w_u, z_v\})$ contains z_u, which is a contradiction. Therefore, a belongs to V_{uv} for some edge uv of H.

If $v \in C$, then, by the definition of S', $u \notin C$, and Claim A implies that S contains all vertices of G_{uv}, which is a contradiction. Hence, $u, v \notin C$.

Let w be some vertex in C. Then $\mathrm{cong_g}(\{a, z_w\})$ contains z_u or z_v, which is a contradiction. This concludes the proof of the claim. □

Together with the previous observations to Claim A, we obtain that S contains at most $2t + (n+5)\binom{t}{2} + n + 1 < k'$ elements, which is a contradiction. This completes the proof of the theorem. □

Now we deal with the calculation of the geodesic convexity number of cographs. We will see that, in this particular case, the problem can be solved in polynomial time.

Theorem 6.8 (Dourado et al. 2012) *Let G be a cograph of order n.*

(*i*) *If G is connected, then*

$$\mathrm{cong_g}(G) = \begin{cases} n-1 & \text{, if } k = 0, \\ \mathrm{cong_g}(G_1) + t - 1 & \text{, if } k = 1, \text{ and} \\ \omega & \text{, if } k \geq 2, \end{cases}$$

where $G_1, \ldots, G_k, \ldots, G_t$ are the connected components of the complement of G, ordered descending by the number of vertices, and k is such that $|V(G_i)| \geq 2$ if and only if $i \leq k$.

(*ii*) *If G is disconnected, then*

$$\mathrm{cong_g}(G) = n - \min\left\{|V(H)| - \mathrm{cong_g}(H) \mid H \text{ is a connected comp. of } G\right\}.$$

Proof

(*i*) If $k = 0$, then G is a complete graph and $\text{cong}_g(G) = n - 1$. If $k = 1$, each vertex u in $G_2, \ldots, G_t$ is adjacent to all vertices in $V(G) \setminus \{u\}$. Let S be a g-convex set of vertices of cardinality $\text{cong}_g(G)$. Let S_1 be the intersection of S and the set of vertices of G_1. Clearly, S_1 is a g-convex set in G_1. If S_1 is a clique, then S_1 does not contain all the vertices of the graph G_1, because G_1 is not complete. By the choice of S, S contains all the vertices in $G_2, \ldots, G_t$. If S_1 is not a clique, then S contains all the vertices in $G_2, \ldots, G_t$, because S is convex. Therefore, $\text{cong}_g(G) = \text{cong}_g(G_1) + t - 1$.

Finally, suppose $k \geq 2$. Let S be a g-convex set of vertices of cardinality $\text{cong}_g(G)$. If S contains two nonadjacent vertices of some G_j, then S contains all the vertices of G outside G_j. Therefore, S contains two nonadjacent vertices outside G_j, which implies that S contains all the vertices of G_j, i.e., S contains all the vertices of G, which is a contradiction. Therefore, S is complete and $\text{cong}_g(G) = \omega$.

(*ii*) It follows directly from the fact that a g-convex set of vertices of G of cardinality $\text{cong}_g(G)$ contains all but one of the connected components of G.

□

Using Theorem 6.8 and the modular decomposition, one can easily calculate the geodesic convexity number of a cograph in linear time.

Still on the geodesic convexity number, Canoy and Garces (2002) study this parameter for graphs obtained from various graph operations. Kim (2004) presents bounds for regular graphs. Chartrand and Zhang (1999) show a Nordhaus and Gaddum (1956)-type inequality for the geodesic convexity number, by showing that $\text{cong}_g(G) + \text{cong}_g(\overline{G}) \leq 2(n(G) - 1)$, and characterize the graphs that achieve equality. This result is deepened by Gimbel (2003), who show an asymptotically optimal lower bound for $\text{cong}_g(G)+\text{cong}_g(\overline{G})$, as well as an NP-hardness reduction for deciding if $\text{cong}_g(G) \geq k$. Gimbel (2003) also presents a Ramsey-type bound for this parameter, by showing bounds for the smallest number of vertices of a graph G such that $\text{cong}_g(G) \geq i$ or $\text{cong}_g(\overline{G}) \geq j$ for given integers i, j. There are bounds for the geodesic convexity number presented in Dourado et al. (2012); Padmavathi (2015). More recently, there are works on the geodesic convexity number of $(q, q - 4)$-graphs (Dourado et al. 2017) and of complementary prism graphs (Castonguay et al. 2019; Neethu and Chandran 2022).

6.3 Other Parameters

We first present some results for the Carathéodory, Radon and Helly numbers when restricted to geodesic convexity. Examples of determining these parameters can be found in Sect. 3.4.

Geodesic Carathéodory Number

When restricted to geodesic convexity, we say that the *geodesic Carathéodory number* of a graph G, denoted by $\mathrm{cth_g}(G)$, is the smallest integer $r \geq 0$ such that, for every subset $S \subseteq V(G)$ and every vertex $u \in \mathrm{conv_g}(S)$, there exists a subset $F \subseteq S$ with $|F| \leq r$ and $u \in \mathrm{conv_g}(F)$. It is also defined as the size of the largest Carathéodory independent set in the geodesic convexity. Remember that, as defined in Sect. 3.4, a subset $S \subseteq V(G)$ is Carathéodory independent in the geodesic convexity if the set

$$\partial(\mathrm{conv_g}(S)) = \mathrm{conv_g}(S) \setminus \left(\cup_{s \in S} \mathrm{conv_g}(S \setminus \{s\})\right)$$

is non-empty.

In view of such a definition, from the point of view of the study of computational complexity related to the geodesic Carathéodory number, there are two decision problems that are directly related. Unfortunately, both are intractable.

Geodesic Carathéodory Number

Instance: Graph G and integer $k \geq 0$.
Question: $\mathrm{cth_g}(G) \leq k$?

Theorem 6.9 (Dourado et al. 2013c) *The Geodesic Carathéodory Number is* NP-*complete.*

Local Geodesic Carathéodory Number

Instance: Graph G, $U \subseteq V(G)$, $u \in \mathrm{conv_g}(U)$ and integer $k \geq 0$.
Question: Is there $F \subseteq U$ such that $|F| \leq k$ and $u \in \mathrm{conv_g}(F)$?

Theorem 6.10 (Dourado et al. 2013c) *The Local Geodesic Carathéodory Number is* NP-*complete, even if G is bipartite.*

While the proof of Theorem 6.9 is more elaborate, the proof of Theorem 6.10 is simpler, relying heavily on the fact that deciding if $\mathrm{hn_g}(G) \leq k$ is NP-complete even if G is bipartite and has a vertex of degree 1 (Araújo et al. 2013).

On the other hand, the same authors show that if G is a split graph, then $\mathrm{cth_g}(G)$ can be calculated in polynomial time, based on the following result.

Theorem 6.11 (Dourado et al. 2013c) *If G is a split graph, then* $\mathrm{cth}_g(G) \leq 3$.

Proof Let G be a split graph and let C and I be a clique and an independent set of G such that $\{C, I\}$ is a partition of $V(G)$. Let $U \subseteq V(G)$ be a Carathéodory independent subset of G. We will prove that $|U| \leq 3$ and, therefore, $\mathrm{cth}_g(G) \leq 3$, since $\mathrm{cth}_g(G)$ is the cardinality of a maximum Carathéodory independent set of G. Let $v \in \partial(\mathrm{conv}_g(U))$, i.e., $v \in \mathrm{conv}_g(S)$ and $v \notin \cup_{s \in S}\mathrm{conv}_g(S \setminus \{s\})$. Let k be the smallest integer such that $v \in \mathrm{I}_g^k(U)$, i.e., the integer such that $v \in \mathrm{I}_g^k(U) \setminus \mathrm{I}_g^{k-1}(U)$. If $k \leq 1$, then $v \in U$ or v belongs to a shortest u, w-path in G for $u, w \in U$. In the first case, note that $U = \{v\}$, as this is the only case where $v \in S$ and $v \notin \cup_{s \in S}\mathrm{conv}_g(S \setminus \{s\})$. In the second case, we have $U = \{u, v\}$, because this is the only way in which $v \in \mathrm{I}_g(u, v)$ and $v \notin \cup_{s \in S}\mathrm{conv}_g(S \setminus \{s\})$. So we can assume that $k \geq 2$.

Since no vertex from the independent set I belongs to a shortest path between two other vertices of U, the set $\mathrm{conv}_g(U) \setminus U$ does not contain any element of I. This implies that $v \in C$. Since $k \geq 2$, the vertex v belongs to a shortest path between two vertices u_{k-1} and w_k in $\mathrm{I}_g^{k-1}(U)$. Therefore, one of the two vertices, say w_k, belongs to $U \cap I$. Furthermore, by the minimality of k, the other vertex u_{k-1} belongs to $\mathrm{I}_g^{k-1}(U) \setminus \mathrm{I}_g^{k-2}(U)$ and, therefore, $u_{k-1} \in C$. Similarly, since $k \geq 2$, the vertex u_{k-1} belongs to a shortest path between two vertices u_{k-2} and w_{k-1} in $\mathrm{I}_g^{k-2}(U)$. As above, we can assume that $w_{k-1} \in U \cap I$. If $k \geq 3$, then this implies that u_{k-2} belongs to $\mathrm{I}_g^{k-2}(U) \setminus \mathrm{I}_g^{k-3}(G)$. Repeating this argument, we obtain vertices $u_0, \ldots, u_{k-1}$ and $w_1, \ldots, w_k$ satisfying:

- $u_0 \in U, u_1, \ldots, u_{k-1} \in C, w_1, \ldots, w_k \in U \cap I$.
- $u_i \in \mathrm{I}_g(\{u_{i-1}, w_i\})$ and $u_i \in \mathrm{I}_g^i(U) \setminus \mathrm{I}_g^{i-1}(U)$, for each $i \in \{1, \ldots, k-1\}$.

Since $v \in \mathrm{conv}_g(\{u_0, w_1, \ldots, w_k\})$ and we want to prove that $|U| \leq 3$, we can assume that $k \geq 3$. By the minimality of k, the vertex u_i is a neighbor of w_k for $i \in \{1, \ldots, k-2\}$; otherwise, $v \in \{u_i, w_k\}$ and, therefore, $v \in \mathrm{I}_g^{i+1}(U)$, a contradiction. Similarly, the vertex u_i is a neighbor of w_{k-1} for all $i \in \{1, \ldots, k-3\}$, if not $u_{k-1} \in \mathrm{I}_g(\{u_i, w_{k-1}\})$ and, then, $v \in \mathrm{I}_g^{i+2}(U)$, a contradiction.

If $k = 3$, then $u_1 \in \mathrm{I}_g(\{w_1, w_3\})$, $u_2 \in \mathrm{I}_g(\{u_1, w_2\})$ and $v \in \mathrm{I}_g(\{u_2, w_3\})$ and, then, $|U| = 3$. If $k \geq 4$, then note that $u_{k-3} \in \mathrm{I}_g(\{w_{k-1}, w_k\})$, $u_{k-2} \in (\{u_{k-3}, w_{k-2}\})$, $u_{k-1} \in \mathrm{I}_g(\{u_{k-2}, w_{k-1}\})$ and $v \in \mathrm{I}_g(\{u_{k-1}, w_k\})$. From this fact, it is possible to deduce that $v \in \mathrm{conv}_g(\{w_{k-2}, w_{k-1}, w_k\})$ and, therefore, $|U| = 3$.

□

In (Lira 2016), the geodesic Carathéodory number of various classes of particular graphs, such as trees and cographs, and complementary prisms of some families of simple graphs, in addition to the Cartesian product of some classes of simple graphs, are studied. Dourado et al. (2017) show that, for a fixed integer $q \geq 4$, $\mathrm{cth}_g(G)$ can be calculated in polynomial time if G is a $(q, q-4)$-graph. More recently, Anand et al. (2020) show polynomial algorithms to calculate the geodesic Carathéodory number for interval graphs and for powers of paths.

Geodesic Radon Number

A set $S \subseteq V(G)$ is *Radon dependent* in the geodesic convexity if there is a partition of S into two sets S_1 and S_2 satisfying $\text{conv}_g(S_1) \cap \text{conv}_g(S_2) \neq \emptyset$ and *Radon-independent* otherwise. The *Radon number* $\text{rd}_g(G)$ of the geodesic convexity is the size of the largest Radon-independent set.

As mentioned in Sect. 3.4, every clique is Radon independent in the geodesic convexity. Therefore, $\text{rd}_g(G) \geq \omega(G)$.

Determining $\text{rd}_g(G)$ is not only an NP-hard problem, but, unless P = NP, there is no approximative algorithm with factor $n(G)^{1-\epsilon}$, for all $\epsilon > 0$, that determines $h\ell_g(G)$, even if G is bipartite (Dourado and da Silva 2017). The geodesic Radon number of d-dimensional grids G was studied by Dourado et al. (2013b), where the authors show bounds, and by Dourado et al. (2016a), who present a polynomial algorithm to determine $\text{rd}_g(G)$. Dourado et al. (2017) show that, for a fixed integer $q \geq 4$, $\text{rd}_g(G)$ can be calculated in polynomial time if G is a $(q, q-4)$-graph, as previously mentioned for the geodesic Carathéodory number. Moran and Yehudayoff (2020) relate bounds on the geodesic Radon number to the existence of ϵ-weak nets.

Geodesic Helly Number

The *Helly number* of the geodesic convexity, denoted by $h\ell_g(G)$, is the size of the largest *Helly independent* subset $S \subseteq V(G)$, which are the sets such that $\bigcap_{v \in S} \text{conv}_g(S \setminus \{v\}) = \emptyset$. Equivalently, if $h\ell_g(G) \geq 2$, then $h\ell_g(G)$ is the smallest integer h such that if $\mathcal{F}$ is the family of g-convex subsets of $V(G)$ and for each subfamily $\mathcal{F}'$ with h members of $\mathcal{F}$ we have that $\bigcap \mathcal{F}' \neq \emptyset$, then $\bigcap \mathcal{F} \neq \emptyset$. Remember that Theorem 3.2 shows that $h\ell_g(G) \geq 2$ for every graph G with at least two vertices.

As mentioned in Sect. 3.4, the Helly number in the monophonic convexity is equal to $\omega(G)$, for every graph G (see Theorem 3.1). From this fact, it follows that $h\ell_g(G) = \omega(G)$ for every distance–hereditary graph G by Lemma 2.3. As noted by Bandelt and Mulder (1990), it is not difficult to construct graphs where $h\ell_g(G) > \omega(G)$ (see Exercise 6.8).

Much of the literature on the geodesic Helly number is focused in determining graph classes for which $h\ell_g(G) = \omega(G)$. Cepoǐ (1986) proved that such equality is valid for chordal graphs using the simplicial elimination scheme. Bandelt and Mulder (1990) generalize the aforementioned results for distance–hereditary graphs and chordal graphs, by showing that such equality also holds for the graph classes *dismantlable* and *pseudo-modular*. Bandelt and Chepoi (1996) generalize these results by proving the same equality in the context of discrete weakly modular spaces. In a series of papers, Polat generalizes the ideas of Bandelt and Mulder (1990) to other classes of finite and infinite graphs (Polat 1995, 2000, 2003).

More recently, computational aspects of determining the geodesic Helly number have been addressed. As mentioned in Theorem 5.3, determining the geodesic Helly number of a graph is an NP-hard problem. More than that, unless P $=$ NP, there is no approximation algorithm with factor $n(G)^{1-\epsilon}$, for every $\epsilon > 0$, that determines $h\ell_g(G)$, even if G is bipartite (Dourado and da Silva 2017). da Silva (2014) implemented an algorithm to determine the geodesic Helly number of graph G and also returns a certificate that shows such graph is not $(h\ell_g(G) - 1)$-Helly. In his dissertation, da Silva (2014) also presents a tight bound for $h\ell_g(G)$, when G is bipartite, in addition to bounds and exact values for certain classes of restricted graphs.

In Carvalho (2016), some particular graph classes also have their geodesic Helly number determined, such as trees, cycles, complete k-partite graphs, complete grids of dimension d, and prism graphs, in addition to a characterization for complete graphs. The author also presents a lower and upper bound for $hn_g(G)$ and shows that deciding whether a graph is p-Helly is co- NP-complete. Finally, he presents ways to calculate the geodesic Helly number of any graph from the determination of the parameter for certain subgraphs.

General Position Number and Rank

As seen in Sect. 3.5, the general position number was introduced in the geodesic convexity, motivated by the *No-Three-in-Line* problem of Dudeney (1917). Manuel and Klavžar (2018) proved that this parameter is NP-complete. A simpler proof can be seen in Theorem 5.3.

The graphs with geodesic general position number 2, $n - 1$ and n were characterized in Thomas and Chandran (2020). The geodesic general position number was studied in Kneser graphs (Patkós 2020; Ghorbani et al. 2021), in cacti and wheel graphs (Yao et al. 2022) and in Cartesian products of simple graphs (Tian and Xu 2021; Tian and Klavžar 2021).

The geodesic rank was introduced by Jamison (1981) and proved NP-hard in Kanté et al. (2017).

Iteration and Percolation Times

The iteration time was introduced by Harary and Nieminem (1981) in the geodesic convexity. In Dourado et al. (2016b), a polynomial algorithm to compute a set with maximum geodesic iteration time was obtained for distance–hereditary graphs. Moscarini (2020) extended this result with a $O(n^3m)$ time algorithm for distance–hereditary graphs and a $O(n^2m)$ time algorithm for bipartite distance–hereditary graphs.

The percolation time was proposed by Béla Bollobás in the P_3 convexity on square grids, a problem that was solved by Benevides and Przykucki (2013). In the geodesic convexity, Benevides et al. (2016) obtained a polynomial algorithm on distance–hereditary graphs and proved that it is NP-complete to decide if $\mathrm{tp_g}(G) \geq 2$ even for bipartite graphs.

Exercises

Exercise 6.1 Present an algorithm in pseudocode with time complexity $O(|S| \cdot m(G))$ such that, given a graph G and a subset $S \subseteq V(G)$, returns $\mathrm{conv_g}(S)$.

Exercise 6.2 Prove Theorem 6.1.

Exercise 6.3 Given integers α and β such that $2 \leq \alpha \leq \beta$, find a graph with $\mathrm{hn_g}(G) = \alpha$ and $\mathrm{in_g}(G) = \beta$.

Exercise 6.4 Show that the **Geodesic Interval Number** belongs to NP.

Exercise 6.5 Show that the graph G' constructed in the proof of Theorem 6.4 is chordal.

Exercise 6.6 Let G be a non-complete graph, S a hull set or an interval set of minimum cardinality in the geodesic convexity of G, and v a universal vertex of G. Show that $v \notin S$.

Exercise 6.7 Prove Theorem 6.6.

Exercise 6.8 Let G be the graph obtained from the complete graph K_n with n vertices by subdividing each edge once. Show that $2 = \omega(G) < \mathrm{h\ell_g}(G) = n$.

Chapter 7
Other Convexities

In this chapter, we present some results of other widely studied graph convexities.

7.1 Monophonic Convexity

We recall that a set of vertices S of a graph G is convex in the monophonic convexity (or m-convex) if S contains all vertices in some induced path between two vertices of S. Note that a clique is an m-convex set.

Monophonic convexity was introduced by Jamison (1982) and several theoretical results were obtained in Jamison and Nowakowski (1984), Farber and Jamison (1986), Duchet (1988). For example, as seen in Sect. 3.4, Jamison and Nowakowski (1984) and Duchet (1988) proved that the monophonic Helly number is always equal to the size of the largest clique in any graph.

Regarding computational complexity, Dourado et al. (2010) proved that the interval number and convexity number are NP-hard in the monophonic convexity. In the same work, the authors presented an algorithm of complexity $O(n^3m)$ to compute the hull number in this convexity for general graphs. Using the results of Leimer (1993), Benevides et al. (2016) adapted this algorithm reducing its complexity to $O(nm)$. Costa et al. (2015) proved that deciding whether the interval number is at most 2 and deciding whether the percolation time is at most 1 are NP-complete problems. They also proved that the convexity number is W[1]-hard and $n^{1-\varepsilon}$-inapproximable for all $\varepsilon > 0$, but polynomial in perfect graphs and planar graphs.

The rank was studied in Dourado et al. (2022b), where the authors showed that this parameter can be computed in polynomial time for the classes of bipartite, cactus, triangle-free graphs, line graphs, and split graphs, this problem being NP-complete for general graphs.

Below we present a characterization of m-convex sets.

J. Araújo et al., *Introduction to Graph Convexity*, SBMAC Collection on Applied and Computational Mathematics, https://doi.org/10.1007/978-3-031-84128-6_7

Theorem 7.1 (Dourado et al. 2010) *Let G be a graph. A subset $S \subseteq V(G)$ is m-convex if and only if for every pair of nonadjacent vertices $u, v \in S$ and every connected component C of $G - S$, we have $V(C) \cap N(u) = \emptyset$ or $V(C) \cap N(v) = \emptyset$.*

Proof Assume that S is m-convex. The existence of a pair of nonadjacent vertices $u, v \in S$ and a connected component C of $G - S$, for which there exist vertices u', v' such that $u' \in V(C) \cap N(u)$ and $v' \in V(C) \cap N(v)$, implies the existence of a sequence of vertices $w_0 = u, w_1 = u', w_2, \ldots, w_{k-1}, w_k = v', w_{k+1} = v$ with $k \geq 1$ satisfying: (i) $w_i \notin S$, $1 \leq i \leq k$; (ii) $u' = v'$ or $(w_i, w_{i+1}) \in E(G - S)$, $1 \leq i \leq k$. Therefore, there exists an induced path connecting u and v which contains at least one vertex outside S, which is a contradiction.

Now consider that S is not m-convex. Let $w_0 = u, w_1, \ldots, w_k, w_{k+1} = v$ be an induced path connecting vertices $u, v \in S$ such that $k \geq 1$ and $w_i \notin S$ for some $i \in \{1, \ldots, k\}$. Let j be an index such that $w_{j-1} \in S$ and $w_j, w_{j+1}, \ldots, w_i \in V(G) \setminus S$. Such an index exists since $u \in S$. Similarly, let ℓ be an index such that $w_i, w_{i+1}, \ldots, w_\ell \in V(G) \setminus S$ and $w_{\ell+1} \in S$. This implies that $w_{j-1}, w_{\ell+1}$ is a pair of nonadjacent vertices of S and there is a connected component C of $G - S$ such that $V(C) \cap N(w_{j-1}) \neq \emptyset$ and $V(C) \cap N(w_{\ell+1}) \neq \emptyset$. □

We use the above result to show that we can answer in polynomial time whether a set is m-convex or not.

Corollary 7.1 (Dourado et al. 2010) *Let G be a graph. Deciding whether a set $S \subseteq V(G)$ is m-convex can be done in time $O(nm)$.*

Proof We describe an algorithm to decide whether $S \subseteq V(G)$ is m-convex with complexity $O(mn)$. Computing the connected components of $G - S$ can be done in time $O(n + m)$. In the worst case, we will have $O(n)$ connected components, say $C_1, \ldots, C_k$. Define the sets $C'_1, \ldots, C'_k$ initially empty. Then, for each edge $uw \in E(G)$ such that $u \in S$ and $w \notin S$, include u in the set C'_i if $w \in C_i$. That is, C'_i is the subset of S formed by the vertices that have at least one neighbor in C_i. If C'_i contains two nonadjacent vertices, stop answering that S is not m-convex. If this process ends and every C'_j, $1 \leq j \leq k$, is a clique, then answer that S is m-convex. □

Now, we show that the decision problem of the MONOPHONIC CONVEXITY NUMBER is NP-complete. In this problem, the instance is a graph G and a positive integer k, and the question is: Is there a subset S of $V(G)$ with at least k vertices that is an interval set in the monophonic convexity?

Theorem 7.2 (Dourado et al. 2010) *The* MONOPHONIC CONVEXITY NUMBER *is* NP-*complete.*

Proof The problem is in NP because testing whether a subset $S \subseteq V(G)$ with $|V(G)| > |S| \geq k$ is m-convex can be done in polynomial time by Corollary 7.1.

To show that this problem is NP-hard, we present a reduction from the CLIQUE problem (Karp 1972): given a graph H and a positive integer ℓ, decide whether H contains a clique of size at least ℓ. We can assume that $\ell < |V(H)| - 1$. Construct a graph G from H as follows. Define $V(G) = V(H) \cup \{u, v\}$, where u and v are new vertices, and $E(G) = E(H) \cup \{(u, x), (v, x) \mid x \in V(H)\}$. Also define $k = \ell + 1$.

We will show that H has a clique with at least ℓ vertices if and only if G has an m-convex set with at least $k = \ell + 1$ vertices.

If $S \subseteq V(H)$ is a clique of size at least ℓ, then $Y = S \cup \{u\}$ is a clique of G and, therefore, is an m-convex set of size at least $\ell + 1$.

Conversely, let Y be a proper subset of $V(G)$ that is m-convex of size at least $\ell + 1$. Note that Y cannot contain both u and v, because if it did, we would have $Y = V(G)$. This means that Y does not contain two nonadjacent vertices w_1 and w_2, as this would imply that u and v belonged to Y. Therefore, $Y \setminus \{u, v\}$ is a clique of size at least ℓ in H. □

A graph G is *split* if $V(G)$ can be partitioned into a clique C and an independent set I. Next, we show how to find in polynomial time the rank of a *split* graph in the monophonic convexity. It is known that this problem is NP-complete for general graphs (Dourado et al. 2022b).

Theorem 7.3 (Dourado et al. 2022b) *The rank of a* split *graph can be found in polynomial time in the monophonic convexity.*

Proof Since Independent Set (Garey and Johnson 1979) belongs to P for bipartite graphs, it suffices to show a polynomial reduction from Rank in the monophonic convexity restricted to *split* graphs to Independent Set restricted to bipartite graphs.

Let G be a split graph with bipartition (C, I). We can assume that C is a maximum clique. Construct a bipartite graph G', from a copy of G, by removing all edges between vertices of C. Denote by (C', I') the bipartition of G' where $C' = \{v_i' : v_i \in C\}$. See Fig. 7.1. We will show that G has an m-convexly

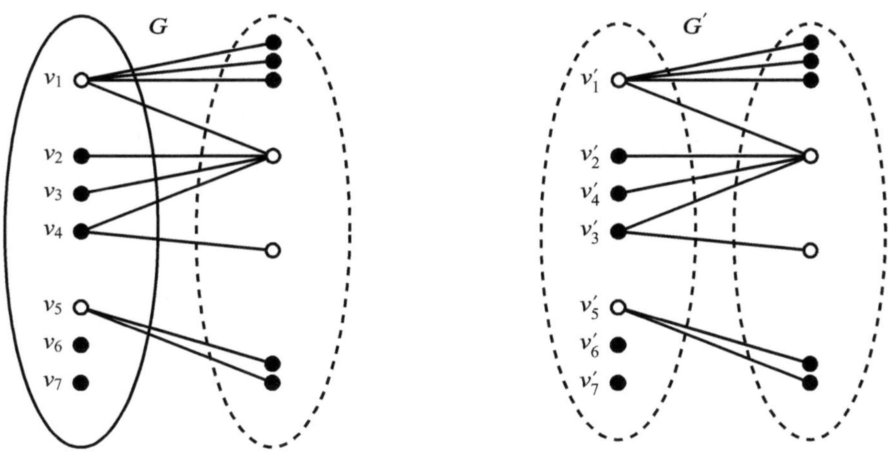

Fig. 7.1 Reduction from the Rank problem in the monophonic convexity in split graphs to Independent Set in bipartite graphs. The vertices in the non-dashed ellipses form a clique

independent set of size at least k if and only if G' has an independent set of size at least k.

First, let S be an m-convexly independent set of size at least k of G. If $|S| \leq |C|$, then G' has an independent set with at least k vertices contained in C' because C' is an independent set. Then, consider $|S| > |C|$ and define $S' = \{v_i' : v_i \in S\}$. If S' is an independent set, then S' is the desired set. Then consider that $v_i'v_j' \in E(G')$ for $v_i', v_j' \in S'$. By the construction of G', without loss of generality, we can assume that $v_j \in C$ and $v_i \in I$. Since S is m-convexly independent, v_i is adjacent to all vertices of $S \cap C$. Since $|S| > |C|$ and C is a maximum clique, then there exists $v_\ell \in C \setminus S$, which implies that $(S \cap I) \setminus \{v_i\} \neq \emptyset$. We also know that every vertex of $(S \cap I) \setminus \{v_i\}$ has no neighbors in $(S \cap C) \cup \{v_t\}$. Then $(S' \setminus \{v_i\}) \cup \{v_\ell\}$ is an independent set of size $|S|$ as desired.

Conversely, let S' be an independent set of G' with at least k vertices. Define $S = \{v_i : v_i' \in S'\}$. Observe that

$$\mathrm{conv}(S \setminus \{v_i\}) \quad \subseteq \quad (S \setminus \{v_i\}) \bigcup \left(\bigcup_{v_j \in (S \setminus \{v_i\}) \cap I} N(v_j) \right).$$

Since S' is an independent set, then $v_i \notin \mathrm{conv}(S \setminus \{v_i\})$, which means that S is an m-convexly independent set of G. □

7.2 Triangle-Path Convexity

In a *triangle path* $v_1, \ldots, v_t$ of a graph G there are no edges connecting vertices v_i and v_j such that $|j - i| > 2$. Thus, a set of vertices $S \subseteq V(G)$ is convex in the *triangle-path convexity* (or tp-convex) if S contains all the vertices that belong to at least one triangle path between two vertices of S.

Note that $\mathrm{conv_g}(S) \subseteq \mathrm{conv_m}(S) \subseteq \mathrm{conv}_t(S)$ and $\mathrm{conv_{p3}}(S) \subseteq \mathrm{conv}_t(S)$, which implies that every tp-convex set is also g-convex, m-convex, and P_3-convex. This convexity was introduced in Bandelt (1989), where some results related to the notion of semispace were obtained. In Changat and Mathews (1999), the values of the Carathéodory, Helly, and Radon numbers in this convexity for general graphs were determined. This convexity was also considered in Changat et al. (2005, 2009).

Dourado and Sampaio (2016) presented polynomial algorithms to find the convexity number and the hull number of general graphs. It is worth noting that, for many convexities, the problems of determining the hull number and the convexity number are NP-complete for general graphs. In the same work, the authors showed that determining the interval number is NP-complete for this convexity even for bipartite graphs. This work also contains a characterization of the tp-convex sets that leads to a polynomial algorithm to recognize such sets. We present this characterization and its consequences below.

Theorem 7.4 (Dourado and Sampaio 2016) *A set of vertices S of a graph G is* tp-*convex if and only if there is no vertex outside S having two neighbors in S and there are no two nonadjacent vertices of S that both have neighbors in one of the connected component of $G - S$.*

Proof Let $S \subset V(G)$ be a tp-convex set. If there is a vertex $v \notin S$ having neighbors $u, w \in S$, then uvw is a triangle path of G connecting two vertices of S such that not all vertices are in S, which would imply that S is not tp-convex. Now consider that there are vertices $u, v \in S$ and $u', v' \notin S$ such that $uv \notin E(G)$, $u' \in N_G(u)$, $v' \in N_G(v)$, and $u', v' \in V(C)$ for some connected component C of $G - S$. If $u' = v'$, then $uu'v$ is a triangle path as in the first case. So we can assume that $u' \neq v'$ and u and v do not have common neighbors in C. Furthermore, without loss of generality, we can assume that u' and v' can be chosen to minimize the distance between them. Now let $P = u'w_1 \dots w_k v'$ for $k \geq 0$ be a shortest path in C. By the choices of u', v' and P, no internal vertex of P is a neighbor of u or v. Then $uu'w_1 \dots w_k v'v$ is an induced path and, thus, a triangle path of G, implying that S is not tp-convex.

Now assume that S is not a tp-convex set. Then there is a triangle path $uw_1 \dots w_k v$ connecting the vertices $u, v \in S$ such that $k \geq 1$ and $w_i \notin S$ for all $i \in \{1, \dots, k\}$. If $uv \in E(G)$, then $k = 1$ and w_1 is a vertex outside S that has two neighbors in S. Otherwise, as $P' = w_1 \dots w_k$ is a path of $G - S$, all vertices of P' belong to the same connected component of $G - S$. Then S has two nonadjacent vertices, u and v, that have neighbors in the same connected component of $G - S$. □

An interesting consequence of the above result is the characterization of the tp-convex sets in terms of the m-convex sets and P_3-convex sets.

Corollary 7.2 (Dourado and Sampaio 2016) *A set of vertices S of a graph G is* tp-*convex if and only if S is* m-*convex and P_3-convex.*

Proof As noted above, the definitions imply that every tp-convex set is m-convex and also P_3-convex. Now consider that S is an m-convex and P_3-convex set and suppose, by contradiction, that S is not tp-convex. The fact that S is P_3-convex implies that there is no vertex outside S having two neighbors in S. As S is not tp-convex, by Theorem 7.4, there are two nonadjacent vertices of S that have neighbors in the same connected component of $G - S$. However, as S is m-convex, Theorem 7.1 implies that there are no two nonadjacent vertices of S that both have neighbors in one of the connected component of $G - S$, which is a contradiction. □

Using these results, we can test whether a set is tp-convex in polynomial time.

Corollary 7.3 (Dourado and Sampaio 2016) *We can test whether a set of vertices S of a graph G of order n and size m is* tp-*convex in $O(nm)$ steps.*

Proof By Corollary 7.2, it is sufficient to test whether S is P_3-convex and m-convex. We leave as an exercise the proof that testing whether S is P_3-convex can be done in time $O(n^2)$. To complete the proof, just apply Corollary 7.1. □

7.3 All-Path Convexity

A set of vertices $S \subseteq V(G)$ is convex in the *all-path convexity* (or ap-convex) if S contains all vertices in some path between two vertices of S. All-path convexity was considered in a series of papers (Sampathkumar 1984; Changat et al. 2001; Gutin and Yeo 2009; Protti and Thompson 2023).

To state the results of this section, we need some definitions. Consider the block decomposition of a graph G, represented by the *block-cut tree* T_G. This tree is defined as follows: each vertex of T_G is associated with a block B_j (a cut edge or a maximal 2-connected subgraph of G) or a cut vertex $z_i \in V(G)$. In addition, there is an edge linking a vertex B_j to a vertex z_i in T_G whenever the block B_j contains the cut vertex $z_i \in V(G)$. This definition implies that the vertices of T_G associated with blocks of G form an independent set, and the same occurs for the vertices of T_G associated with cut vertices of G (Exercise 7.1). Moreover, each leaf of T_G represents a block of G. A *terminal block* of G is a block associated with a leaf of T_G.

For a set $S \subseteq V$, let T_S be the maximal subtree of T_G such that each leaf of T_S is associated with a block of G containing a vertex of S that is not a cut vertex in the subgraph G_S induced by $\cup_{B_j \in V(T_S)} B_j$. Figure 7.2 below shows an example.

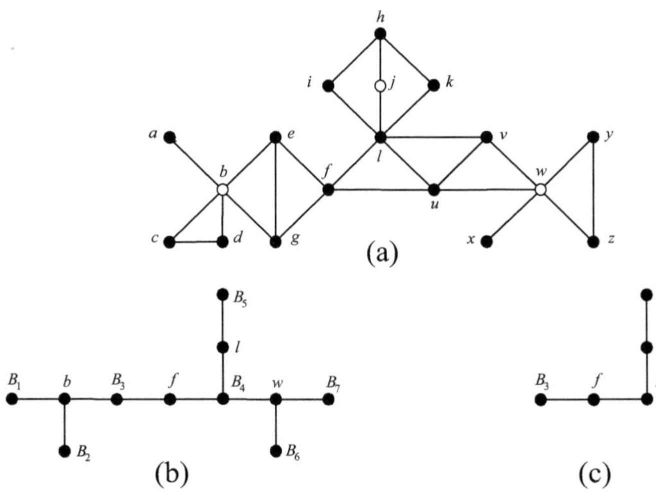

Fig. 7.2 (a) A graph G and a subset $S = \{b, j, w\}$ (represented by the white vertices), whose blocks are such that $V(B_1) = \{a, b\}$, $V(B_2) = \{b, c, d\}$, $V(B_3) = \{b, e, g, f\}$, $V(B_4) = \{f, l, u, v, w\}$, $V(B_5) = \{h, i, j, k, l\}$, $V(B_6) = \{w, x\}$, $V(B_7) = \{w, y, z\}$; (b) block-cut tree T_G; (c) subtree T_S of T_G. The block B_3 is a leaf of T_S because it does not contain any cut vertex in the graph G_S induced by $V(B_3) \cup V(B_4) \cup V(B_5)$

We will also need the following lemma, presented without proof:

Lemma 7.1 (Protti and Thompson 2023) *Let $S \subseteq V$ and u, w be two distinct vertices in S, belonging to the blocks B_u and B_w of T_S, respectively. Assume that u and w are not cut vertices in G_S. Let $B_{j_1} z_1 B_{j_2} z_2 \ldots z_{k-1} B_{j_k}$ be a path in T_S between $B_{j_1} = B_u$ and $B_{j_k} = B_w$. Then, for each $v \in \cup_{i=1}^{k} V(B_{j_i})$, there exists a path P in G from u to w passing through v.*

Let us now focus on the problem of determining the convexity number of a graph G in the all-path convexity, denoted by $\mathrm{con}_{\mathrm{ap}}(G)$.

For a terminal block B_j of G, let $b_j = |V(B_j)|$. Furthermore, define $b(G) = \min\{b_j \mid B_j \text{ is a terminal block of } G\}$.

Theorem 7.5 (Protti and Thompson 2023) *For any graph G, it holds that*

$$\mathrm{con}_{\mathrm{ap}}(G) = \begin{cases} 1, & \textit{if } |V(G)| = 2 \textit{ or } G \textit{ is 2-connected;} \\ n - b(G) + 1, & \textit{otherwise.} \end{cases}$$

Proof If $|V(G)| = 2$, then the theorem is trivially true. If G is 2-connected, then, by the Fan Lemma (see Proposition 9.5 in (Bondy and Murty 2008)), for each pair of vertices $u, w \in V$, $w \neq u$, every $v \notin \{u, w\}$ is on a path from u to w. Therefore, for every S with $2 \leq |S| \leq n-1$, S is not convex. This implies that $\mathrm{con}_{\mathrm{ap}}(G) = 1$.

Now suppose that G is not 2-connected. Note that any $S \subseteq V(G)$ that consists of the union of all vertex sets of all blocks of G, except one terminal block, say B_j, is convex, because the only cut vertex z belonging to $V(B_j)$ separates all vertices of $V \setminus V(B_j)$ from $V(B_j) \setminus \{z\}$ (note that $z \in S$). Thus, the maximum convex set in G is obtained by removing from G all vertices in a terminal block B_j of minimum size, except the cut vertex $z \in V(B_j)$. □

Let us now consider the problems of determining the interval number and the hull number of G in the all-path convexity. These parameters are denoted by $\mathrm{in}_{\mathrm{ap}}(G)$ and $\mathrm{hn}_{\mathrm{ap}}(G)$, respectively. Let $eb(G)$ be the number of terminal blocks of G.

Theorem 7.6 (Protti and Thompson 2023) *For any graph G, it holds that*

$$\mathrm{in}_{\mathrm{ap}}(G) = \begin{cases} 1, & \textit{if } G \textit{ is trivial;} \\ 2, & \textit{if } |V(G)| = 2 \textit{ or } G \textit{ is 2-connected;} \\ eb(G), & \textit{otherwise.} \end{cases}$$

Proof If $|V(G)| \leq 2$, the theorem is trivially true. If G is 2-connected, by the Fan Lemma any pair $u, w \in V(G)$, $w \neq u$, is such that the interval of $\{u, w\}$ is equal to $V(G)$ and, thus, $\mathrm{in}_{\mathrm{ap}}(G) = 2$ in this case.

Finally, if G is not 2-connected, consider $S \subseteq V(G)$ such that $S \cap V(B_j) = \{v_j\}$ for each terminal block B_j of G, where v_j is not a cut vertex of G. Note that $|S| = eb(G)$. The definition of S implies that $T_S = T_G$ and, thus, every vertex $v \in V(G)$

is in a block B_v of G belonging to a maximum path $B_{j_1} z_1 B_{j_2} z_2 \ldots z_{k-1} B_{j_k}$ in T_S such that B_{j_1} and B_{j_k} are terminal blocks of $G_S = G$, containing vertices $u, w \in S$, $w \neq u$, respectively, that are not cut vertices in $G_S = G$. By Lemma 7.1, there is a path P in G from u to w passing through v. In other words, the interval of S is equal to $V(G)$. To conclude the proof, if a set $S' \subseteq V(G)$ is such that $|S'| < eb(G)$, then there is at least one terminal block B_j in G such that $V(B_j) \setminus \{z_j\}$ does not contain vertices of S', where z_j is the cut vertex of G belonging to $V(B_j)$. Therefore, no vertex in $V(B_j) \setminus \{z_j\}$ can be on a path starting and ending in distinct vertices of S', i.e., the interval of S' is not equal to $V(G)$. Thus, S is minimum. □

It can be shown that the interval of any set $S \subseteq V(G)$ is ap-convex, implying that, in the all-path convexity, the interval and the convex hull of any S coincide. Hence, the following corollary holds:

Corollary 7.4 *For any graph G,* $\text{in}_{\text{ap}}(G) = \text{hn}_{\text{ap}}(G)$.

As a consequence of the results presented in this section, it is easy to check that the parameters $\text{con}_{\text{ap}}(G)$, $\text{in}_{\text{ap}}(G)$, and $\text{hn}_{\text{ap}}(G)$ can be computed in linear time in the size of G.

7.4 Steiner Convexity

Given a connected graph G and a set $S \subseteq V(G)$, let T be a connected subgraph of G with a minimum number of edges that contains all the vertices of S. It is easy to see that T is necessarily a tree (Exercise 7.2), called the *Steiner tree* of S. Finding a Steiner tree of a set S is a problem widely studied in the literature, as it generalizes the concept of minimum path. Note that, if $|S| = 2$, then T is precisely a minimum path between the vertices of S. Thus, $|T|$ is exactly the smallest number of edges necessary to connect all the vertices of the set S in a subgraph.

A set S is said to be St-*convex* if for any $S' \subseteq S$, the vertices of any Steiner tree of S' belong to S. The family of all St-convex sets of a graph G defines a convexity called *Steiner convexity* of G, introduced in Cáceres et al. (2008).

Given $S \subseteq V(G)$, the *Steiner interval* $I_{St}(S)$ of S is defined as follows:

$$I_{St}(S) = \bigcup_{S' \subseteq S} \{V(T_{S'}) \mid T_{S'} \text{ is a Steiner tree of } S'\}.$$

In Dourado et al. (2014), it is proved that, given a vertex x and a set S, determining whether x belongs to $I_{St}(S)$ is an NP-hard problem.

Exercises

Exercise 7.1 Let G be a connected graph. Let T_G be a graph defined as follows: each vertex of T_G is associated with a block B_j (a cut edge or a maximal 2-connected subgraph of G) or a cut vertex $z_i \in V(G)$. In addition, there is an edge linking a vertex B_j to a vertex z_i in T_G whenever the block B_j contains the cut vertex $z_i \in V(G)$. Show that T_G is a tree.

Exercise 7.2 Given a connected graph G and a set $S \subseteq V(G)$, show that a connected subgraph of G with the fewest number of edges that contains all the vertices of S is a tree. In addition, show that every leaf of this tree is a vertex of S.

Chapter 8
Convexity in Oriented Graphs

An *oriented graph* D is an orientation of a simple graph G, i.e., D is obtained from G by assigning a direction for each edge. Alternatively, an oriented graph is a directed graph with no symmetric pair of arcs. Despite the oriented case being less studied in the literature, compared to the undirected case, some of the oldest papers on graph convexity deal exactly with the oriented case (Erdős et al. 1972; Moon 1972). For the oriented case, two convexities have been studied in the literature: the geodesic and the P_3.

In the geodesic convexity, the corresponding interval function $\overrightarrow{\mathrm{I}}_{\mathrm{g}}(u, v)$ returns all the vertices of D that belong to all shortest oriented (u, v)-paths or (v, u)-paths (Chartrand and Zhang 2000; Chartrand et al. 2003). Given the definition of the interval function for an oriented graph, the other definitions and parameters essentially have the same definition as in the undirected case.

Note that, for any two vertices u, v of a g-*convex* set S in an oriented graph D, every vertex w in some shortest (u, v)-path or (v, u)-path in D also belongs to S. When a subset of vertices S of an oriented graph D has $V(D)$ as its hull, denoted by $\overrightarrow{\mathrm{conv}}_{\mathrm{g}}(S) = V(G)$, we say that this set is a *geodesic hull set* of D. The *geodesic hull number* $\overrightarrow{\mathrm{hn}}_{\mathrm{g}}(D)$ of an oriented graph D in the geodesic convexity is the smallest cardinality of a geodesic hull set of D. The *geodesic interval number* of D, denoted by $\overrightarrow{\mathrm{in}}_{\mathrm{g}}(D)$, is the cardinality of a smallest set S such that $\overrightarrow{\mathrm{I}}_{\mathrm{g}}(S) = V(D)$. If S is such that $\overrightarrow{\mathrm{I}}_{\mathrm{g}}(S) = V(D)$, then S is a *geodesic interval set* of D. It should be highlighted that, in the particular case of geodesic convexity, as occurs in the undirected case, the interval number is also known as the *geodesic number* of D.

Figure 8.1 shows an example. Note that u_1, u_2, u_3 are sinks and x_1 is a source and, therefore, they belong to every hull set and every interval set in the geodesic convexity of D. Therefore, $\overrightarrow{\mathrm{hn}}_{\mathrm{g}}(D) \geq 4$. The (x_1, u_i)-shortest paths, with $i \in \{1, 2, 3\}$, cover almost all vertices, except for v_1 and v_2. As (t_j, v_j, z_j) is a shortest path for $j \in \{1, 2\}$, we have that $I^2[\{x_1, u_1, u_2, u_3\}] = V(D)$ and, therefore, $\overrightarrow{\mathrm{hn}}_{\mathrm{g}}(D) = 4$. Next, note that the vertices v_j's must be in a minimum geodesic

J. Araújo et al., *Introduction to Graph Convexity*, SBMAC Collection on Applied and Computational Mathematics, https://doi.org/10.1007/978-3-031-84128-6_8

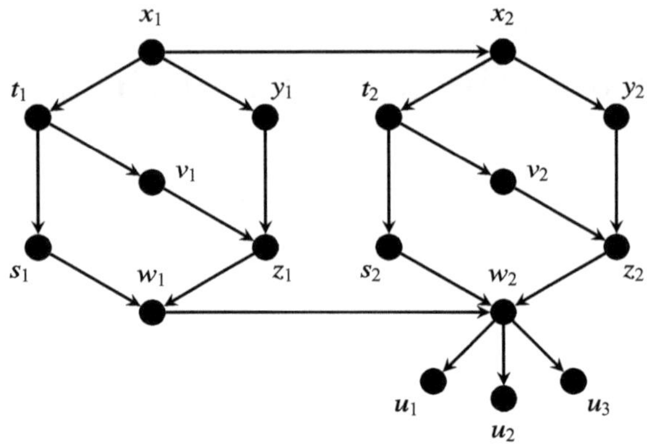

Fig. 8.1 Oriented graph D with $\overrightarrow{\text{hn}}_g(D) = 4$ and $\overrightarrow{\text{in}}_g(D) = 6$

set. Thus, $\{x_1, u_1, u_2, u_3, v_1, v_2\}$ is a minimum geodesic set and, consequently, $\overrightarrow{\text{in}}_g(D) = 6$.

As argued above, sources and sinks are minimum coconvex sets in the geodesic convexity of oriented graphs. They are not the only ones. A vertex v of an oriented graph D is *transitive* if, for any $u, w \in V(D)$ such that $(u, v) \in A(D)$ and $(v, w) \in A(D)$, then $(u, w) \in A(D)$. With this, there is no (x, y)-shortest path in D containing v if $x \neq v$ and $y \neq v$, since one can always take a *shortcut* in the neighborhood of v. The notion of transitive vertex corresponds to that of simplicial vertex in the undirected case. In the literature on geodesic convexity in oriented graphs, a vertex v of an oriented graph D is *extreme* if v is a source, sink, or transitive.

In the convexity $\overrightarrow{P_3}$ of an oriented graph D, the interval function $\overrightarrow{I}_{p3}(u, v)$ returns all vertices of D that belong to directed (u, v)-paths or (v, u)-paths with exactly 3 vertices (Erdős et al. 1972). We use the terms $\overrightarrow{P_3}$*-convex set*, $\overrightarrow{P_3}$ *hull set*, $\overrightarrow{P_3}$ *hull number* $\overrightarrow{\text{hn}}_{p3}(D)$, $\overrightarrow{P_3}$ *interval set*, and $\overrightarrow{P_3}$ *interval number* $\overrightarrow{\text{in}}_{p3}(D)$.

In the convexity $\overrightarrow{P_3}$, a subset S of vertices is convex if, for each $v \in V \setminus S$, all arcs are directed either from S to v or from v to S. Moreover, sources and sinks remain unitary coconvex sets, but the same does not occur for transitive vertices.

Next, we present the state of the art on convexity in oriented graphs with respect to the parameters hull number and interval number and to the related parameters in the geodesic and P_3 convexities, organized by the type of contribution. We should highlight that there are works in the context of oriented graphs on the *convexity number* in the geodesic convexity (Chartrand et al. 2002a) and on the *rank* and the *Carathéodory*, *Radon*, and *Helly* numbers in the $\overrightarrow{P_3}$ convexity (Parker et al. 2006, 2008, 2009; Parker and Westhoff 2012). These latter ones focus mainly on the study of these parameters when restricted to the class of complete multipartite graphs.

Erdős et al. (1972); Moon (1972) study properties of convex sets in the $\overrightarrow{P_3}$ convexity in the class of *tournaments*, shown in the next section.

8.1 The Class of Tournaments

A *tournament* is an orientation of a complete graph. Regarding the $\overrightarrow{P_3}$ convexity, Erdős et al. (1972) define a tournament T as *simple* if the set of nontrivial convex subsets of T is empty. As in the case of undirected graphs, a convex set $S \subseteq V(D)$ of an oriented graph D is *trivial* if $|S| = 1$ or $S = V(D)$. They use the notation $\mathrm{C}(T)$ for the family of nontrivial convex subsets of a tournament T.

Theorem 8.1 (Erdős et al. 1972) *Every tournament T can be extended into a simple tournament T' with two more vertices whenever $|T| \neq 2$.*

Moon (1972) refined Theorem 8.1: except in particular cases, any tournament is contained in a simple tournament with an extra vertex. Erdős et al. (1972) also present lower and upper bounds for the number of simple tournaments.

Theorem 8.2 (Erdős et al. 1972) *If T is a finite tournament, then $|C(T)| \leq \binom{|V(T)|}{2} - 1$ with equality only if T is transitive.*

Theorem 8.3 (Erdős et al. 1972) *If α is an infinite cardinal, then there exist 2^{α} non-isomorphic simple tournaments of order α.*

Finally, Erdős et al. (1972) also show that $C(T)$ satisfies the Bernstein property and the class of simple tournaments is not pseudoelementary.

Still on tournaments, Haglin and Wolf (1996) show that $\overrightarrow{\mathrm{hn}}_{\mathrm{p3}}(T) \leq 2$, which implies, as it was of the authors' interest to find, the existence of an $O(n^4)$ algorithm to determine *all* the convex sets in this class. These results for tournaments, along with the aforementioned works on the rank and the Carathéodory, Radon, and Helly numbers in orientations of complete multipartite graphs, constitute the vast majority of the works found in the literature on the $\overrightarrow{P_3}$ convexity in oriented graphs. In the following sections, we present the other results that are mainly for geodesic convexity.

8.2 Bounds, Properties, and Existential Results

Bounds

It is clear that both parameters $\overrightarrow{\mathrm{in}}_{\mathrm{g}}(D)$ and $\overrightarrow{\mathrm{hn}}_{\mathrm{g}}(D)$ are equal to one when D is an oriented graph with only one vertex. Otherwise, there are no good bounds in general:

Proposition 8.1 (Chartrand and Zhang 2000; Chartrand et al. 2003) *If D is a nontrivial oriented graph, then*

$$2 \leq \overrightarrow{\mathrm{hn}}_{\mathrm{g}}(D) \leq \overrightarrow{\mathrm{in}}_{\mathrm{g}}(D) \leq n.$$

Moreover, these bounds can be reached by both parameters.

Proof See Exercise 8.1. □

There are also upper bounds on the diameter of the oriented graph.

Proposition 8.2 (Chartrand et al. 2003; Chartrand and Zhang 2000) *Let D be a nontrivial oriented graph. The following bounds are tight:*

$$\overrightarrow{\text{hn}}_{\text{g}}(D) \leq \overrightarrow{\text{in}}_{\text{g}}(D) \leq n(D) - \text{diam}(D) + 1.$$

Proof See Exercise 8.2. □

Characterizations

In the previous section, we presented lower and upper bounds for the two parameters. There are also results in the literature characterizing the graph class that achieve such bounds. For the upper bound, we find characterizations for the oriented graphs D such that $\overrightarrow{\text{hn}}_{\text{g}}(D) = n$ and $\overrightarrow{\text{in}}_{\text{g}}(D) = n$. An oriented graph D is *transitive* whenever $(u, w) \in A(D)$ whenever there is $v \in V(D)$ such that $(u, v), (v, w) \in A(D)$.

Proposition 8.3 (Chartrand et al. 2003; Chartrand and Zhang 2000) *If D is a nontrivial oriented graph, then*

1. $\overrightarrow{\text{hn}}_{\text{g}}(D) = n(D)$ *if and only if D is transitive.*
2. $\overrightarrow{\text{in}}_{\text{g}}(D) = n(D)$ *if and only if D is transitive.*

Proof If D is transitive, each vertex $v \in V(D)$, satisfying $d_D^+(v) > 0$ and $d_D^-(v) > 0$, must necessarily be a transitive vertex. All other vertices of D are sources or sinks. Hence, all vertices of D are extremes and, therefore, $\overrightarrow{\text{hn}}_{\text{g}}(D) = \overrightarrow{\text{in}}_{\text{g}}(D) = n$. For the other implication, consider the contrapositive of the same. If D is not transitive, there are vertices $u, v, w \in V(D)$ such that $(u, v), (v, w) \in A(D)$ and $(u, w) \notin A(D)$. In this case, $\overrightarrow{I}_{\text{g}}[\{u, w\}]$ has the vertex v and, thus, $V(D) \setminus \{v\}$ is a set of hull and geodesic of D. Consequently, $\overrightarrow{\text{hn}}_{\text{g}}(D) \leq n(D)-1$ and $\overrightarrow{\text{in}}_{\text{g}}(D) \leq n(D)-1$. □

For the lower bound, unfortunately we did not find a characterization like those above. There is at least the following related result, by Chartrand and Zhang (2000).

Proposition 8.4 (Chartrand and Zhang 2000) *Let D be an oriented graph such that $n(D) \geq 3$. Then every pair of vertices of D is a set of interval in the geodesic convexity of D if and only if D is a directed cycle.*

Existential Results

In this subsection, we will present results on the existence of oriented graphs that obey certain characteristics. The proof of the following proposition is left to Exercise 8.3.

Proposition 8.5 (Chartrand et al. 2003) *For any two integers k and n with $2 \leq k \leq n$, there is an orientation of P_n with order n and hull number k.*

Proposition 8.5 can also be used for the analogous result with the parameter $\overrightarrow{\text{in}}_g$.

Proposition 8.6 (Chartrand and Zhang 2000) *For any two integers k and n with $2 \leq k \leq n$, there is an orientation of P_n with order n and geodesic number k.*

Proof See Exercise 8.3. □

Moreover, Chartrand and Zhang (2000) show that, in the above proposition, we can restrict ourselves to tournaments.

Proposition 8.7 (Chartrand and Zhang 2000) *For any two integers k and n with $2 \leq k \leq n$, there is a tournament of order n and geodesic number k.*

The next result, on the one hand, seems more general than the previous ones. It asserts the existence of an oriented graph D with $\overrightarrow{\text{hn}}_g(D) = a$ and $\overrightarrow{\text{in}}_g(D) = b$ for each pair of integers $2 \leq a \leq b$. The difference is that, unlike Propositions 8.5 and 8.6, the graph shown below has the number of vertices determined by a and b.

Proposition 8.8 (Chartrand et al. 2003) *For any integers $2 \leq a \leq b$, there is a connected oriented graph D such that $\overrightarrow{\text{hn}}_g(D) = a$ and $\overrightarrow{\text{in}}_g(D) = b$.*

The next result is a generalization of Proposition 8.5, since P_n has $n - 1$ edges.

Theorem 8.4 (Chartrand et al. 2003) *For each pair of integers n, m with $n - 1 \leq m \leq \binom{n}{2}$, there is a graph G of order n and size m such that, for each integer k with $2 \leq k \leq n$, there is an orientation D of G such that $\overrightarrow{\text{hn}}_g(D) = k$.*

Chartrand and Zhang (2000) asked if Theorem 8.4 also holds for the geodesic interval number, which was answered by Chang et al. (2004).

Theorem 8.5 (Chang et al. 2004) *For each pair of integers n, m with $n - 1 \leq m \leq \binom{n}{2}$, there is a graph G of order n and size m such that, for each integer k with $2 \leq k \leq n$, there is an orientation D of G such that $\overrightarrow{\text{in}}_g(D) = k$.*

Proposition 8.9 (Chartrand and Zhang 2000) *For every integer k, there exists an oriented graph D and an arc $a \in A(D)$ such that the inversion of the direction of a, producing the oriented graph D', results in $\overrightarrow{\text{in}}_g(D') = \overrightarrow{\text{in}}_g(D) + k$.*

8.3 Geodesic Spectrum

Given a graph G, the *geodesic spectrum* of G, denoted by $S_g(G)$, is the set of geodesic interval number values among all orientations of G. That is, $S_g(G) = \{\overrightarrow{\text{in}}_g(D) \mid D \text{ orientation of } G\}$. Similarly, the *hull spectrum* of a graph G, denoted by $S_h(G)$, is defined with respect to the geodesic hull numbers of orientations D of G. The spectrum will be said to be *continuous* if it corresponds to the set $\{2, \ldots, n\}$. By Theorems 8.4 and 8.5, there exist graphs with continuous hull and geodesic spectra for each triple n, m, k, satisfying the appropriate conditions.

From Propositions Proposition 8.5 and 8.6, we have the following.

Corollary 8.1 *For $n \geq 2$, $S_h(P_n) = S_g(P_n) = \{2, \ldots, n\}$.*

So far, notice that we have only shown graphs with continuous geodesic and hull spectra. However, this is not always the case, as the following result shows.

Theorem 8.6 (Chang et al. 2004) *For every $n \geq 3$,*

$$S_g(C_n) = \{3\} \cup \left\{2s \mid 1 \leq s \leq \left\lfloor \frac{n}{2} \right\rfloor\right\}.$$

Proof See Exercise 8.4. □

The version for the hull spectrum is a corollary of the following result, and the argument is analogous to the previous one.

Proposition 8.10 (Chartrand et al. 2003) *Let D be an orientation of C_n. Then $\overrightarrow{\text{hn}}(D) = 3$ or $\overrightarrow{\text{hn}}(D) = 2t$ for some integer t with $1 \leq t \leq n/2$.*

Corollary 8.2 *For $n \geq 3$, $S_h(C_n) = \{3\} \cup \{2s \mid 1 \leq s \leq \lfloor \frac{n}{2} \rfloor\}$.*

The next result is about the geodesic spectrum of trees. Notice that, as the leaves of a tree T have only one neighbor, in an orientation of T, each leaf will be a source or a sink. Thus, being ℓ the number of leaves of T, it follows that $\overrightarrow{\text{in}}_g(D) \geq \ell$ for every orientation D of T.

Theorem 8.7 (Chang et al. 2004) *If T is a tree with ℓ leaves and n vertices, then $S_g(T) = \{\ell, \ell + 1, \ldots, n\}$.*

By Proposition 8.7, we can deduce that $S_g(K_n) = \{2, \ldots, n\}$. The theorem below generalizes this result for r-partite graphs with minimum degree at least two.

Theorem 8.8 (Chang et al. 2004) *If $G = K_{n_1,\ldots,n_r}$ is a complete r-partite graph of order n and such that $\delta(G) \geq 2$, then $S_g(G) = \{2, \ldots, n\}$.*

8.4 Maximums and Minimums Among All Orientations

Most of the literature on geodesic convexity in oriented graphs deals with the parameters defined below. Let G be an undirected graph. The *upper orientable hull number* and the *lower orientable hull number* are defined respectively by

$$\mathrm{hn}^+(G) = \max\{\overrightarrow{\mathrm{hn}}_{\mathrm{g}}(D) \mid D \text{ is an orientation of } G\};$$

$$\mathrm{hn}^-(G) = \min\{\overrightarrow{\mathrm{hn}}_{\mathrm{g}}(D) \mid D \text{ is an orientation of } G\}.$$

Basically these are the largest and smallest value among the elements of $S_h(G)$, respectively. Notice that, as we are dealing with finite graphs, both numbers are well defined. The *upper orientable geodesic interval number* and the *lower orientable geodesic interval number* are defined in an analogous way by

$$\mathrm{gn}^+(G) = \max\{\overrightarrow{\mathrm{in}}_{\mathrm{g}}(D) \mid D \text{ orientation of } G\};$$

$$\mathrm{gn}^-(G) = \min\{\overrightarrow{\mathrm{in}}_{\mathrm{g}}(D) \mid D \text{ orientation of } G\}.$$

By the lower bound of Proposition 8.1, we have that $\mathrm{hn}^-(G) \geq 2$ and $\mathrm{gn}^-(G) \geq 2$. Thus, a natural question is which graphs reach this bound. Although there are no characterizations, there are sufficient conditions in the literature. See Exercise 8.5.

Proposition 8.11 (Chartrand and Zhang 2000) *Let G be a nontrivial connected graph. If G has a Hamiltonian path, then* $\mathrm{gn}^-(G) = 2$.

Proposition 8.12 (Chartrand et al. 2003) *Let G be a nontrivial connected graph. If G has a Hamiltonian path, then* $\mathrm{hn}^-(G) = 2$.

There is also an upper bound for $\mathrm{hn}^-(G)$ presented in the literature. Given a graph G, a *spanning tree* of G is a tree $T \subseteq G$ such that $V(T) = V(G)$.

Lemma 8.1 (Chartrand et al. 2003) *Let μ be the smallest number of leaves in a spanning tree of the connected graph G; then* $\mathrm{hn}^-(G) \leq \mu$.

Note that this result also holds for the lower orientable geodesic interval number.

Theorem 8.9 (Dong and Wang 2009) *Given any graph G,* $\mathrm{gn}^-(G) \leq \min\{\ell(T) \mid$ *T is a spanning tree of G where $\ell(T)$ is the number of leaves of T.*

Regarding upper bounds, there is the following result in the literature.

Proposition 8.13 (Chartrand et al. 2003) *Let G be a nontrivial connected graph. Then,* $\mathrm{hn}^+(G) = n$ *if and only if G has a transitive orientation if and only if* $\mathrm{gn}^+(G) = n$.

In Chartrand and Zhang (2000), it was also shown that bipartite graphs of order at least two reach this value for the upper orientable geodesic number. Remember that $\overrightarrow{\text{hn}}_{\text{g}}(D) \leq \overrightarrow{\text{in}}_{\text{g}}(D)$ for every oriented graph D. Therefore, it is easy to observe that $\text{hn}^{-}(G) \leq \text{gn}^{-}(G)$, as well as $\text{hn}^{+}(G) \leq \text{gn}^{+}(G)$ for every graph G. Also note that $\text{hn}^{-}(G) \leq \text{hn}^{+}(G)$ for every graph G and that $\text{gn}^{-}(G) \leq \text{gn}^{+}(G)$.

If equality holds, this would imply that the values of the hull number (geodesic) of all possible orientations of a graph are the same. At first glance, it seems strange that if we take a graph of large order and size, all its orientations have the same hull number or the same geodesic number. Indeed, both inequalities are strict.

Theorem 8.10 (Farrugia 2005) *For every connected graph G with at least three vertices, we have* $\text{hn}^{-}(G) < \text{hn}^{+}(G)$ *and* $\text{gn}^{-}(G) < \text{gn}^{+}(G)$.

With this, $\text{hn}^{-}(G) < \text{hn}^{+}(G) \leq \text{gn}^{+}(G)$ and $\text{hn}^{-}(G) \leq \text{gn}^{-}(G) < \text{gn}^{+}(G)$. Next, we present a relationship between $\text{gn}^{-}(G)$ and $\text{hn}^{+}(G)$.

Theorem 8.11 (Hung et al. 2009) *For every connected graph G with at least three vertices, we have* $\text{gn}^{-}(G) < \text{hn}^{+}(G)$. *Consequently,*

$$\text{hn}^{-}(G) \leq \text{gn}^{-}(G) < \text{hn}^{+}(G) \leq \text{gn}^{+}(G).$$

8.5 Complexity

The study of computational complexity to determine the hull number and interval number parameters was only the main focus of just one recent work in the literature. In it, the authors exclusively study the case of geodesic convexity.

Theorem 8.12 (Araújo and Arraes 2022) *Given an oriented partial cube D and an integer* $k \geq 1$, *deciding whether* $\overrightarrow{\text{hn}}_{\text{g}}(D) \leq k$ *is* NP*-complete.*

The general idea of the proof of Theorem 8.12 is simple, but follows from two not necessarily trivial observations: by replacing each edge $uv \in E(G)$ of a nondirected bipartite graph G with a directed C_4, where u and v are nonconsecutive vertices of the C_4, produces an oriented graph D such that $\overrightarrow{\text{hn}}_{\text{g}}(D) = \text{hn}_{\text{g}}(G)$, and if G is a partial cube, then the same operation of replacing each edge with a C_4 produces a partial cube. Therefore, the result follows since determining whether $\text{hn}_{\text{g}}(G) \leq k$, given a nondirected graph G and a positive integer k, is an NP-complete problem, even if G is a partial cube (Albenque and Knauer 2016). It should be noted that the class of partial cubes is a subclass of bipartite.

Araújo and Arraes (2022) also show that determining the geodesic interval number of an oriented graph D is a computationally hard problem. The reduction is made from the set cover problem.

Set Cover

Instance:	Set $U, \mathcal{F} \subseteq \mathcal{P}(U)$ where $\bigcup_{F \in \mathcal{F}} F = U$ and $k \in \mathbb{Z}_+^*$.
Parameter:	k.
Question:	Is there $\mathcal{F}' \subseteq \mathcal{F}$ such that $\bigcup_{F \in \mathcal{F}'} F = U$ and $\lvert\mathcal{F}'\rvert \leq k$?

Set Cover is one of the 21 NP-complete problems of Karp (1972), it does not admit an $O(\log n)$-approximation algorithm, unless P = NP (Lund and Yannakakis 1994), and is W[2]-hard when parameterized by k (Downey and Fellows 2012).

Theorem 8.13 (Araújo and Arraes 2022) *Deciding whether* $\overrightarrow{\text{in}}_g(D) \leq k$ *is* NP-*complete, is* $O(\log n)$*-inapproximable and is* W[2]*-hard when parameterized by k, even if D is an acyclic orientation of a bipartite, cobipartite, or split graph.*

Proof We show the reduction for bipartite graphs. The other two cases follow from small modifications of it and they are left as an exercise (see Exercise 8.6).

First, note that one can calculate in polynomial time $\overrightarrow{I}_g(S)$, for a given candidate solution S in an oriented graph D, as well as in the undirected case as presented in Dourado et al. (2009). Therefore, the problem is NP.

Given an instance $I = (U, \mathcal{F}, k)$ of the **Set Cover**, we will construct a bipartite oriented graph $D(I)$ with bipartition $\{A, B\}$ of $V(D)$ into independent sets, such that I is a YES instance for **Set Cover** if and only if $\overrightarrow{\text{in}}_g(D(I)) \leq k + 3$. Assume, without loss of generality, that $U = \{1, \ldots, n\}$ and that $\lvert\mathcal{F}\rvert = m$.

For each $F_i \in \mathcal{F}$, a corresponding vertex f_i is added in A. For each $j \in U$, a vertex u_j is added in B. Whenever $j \in F_i$, the arc (f_i, u_j) is added to $A(D(I))$ for each $i \in \{1, \ldots, m\}$ and for each $j \in \{1, \ldots, n\}$.

Three more vertices u, v, and w are added, as well as the arcs $(u, f_i), (f_i, w)$, for all $i \in \{1, \ldots, m\}$, (u_j, v) for all $j \in \{1, \ldots, n\}$ and finally (u, v). v is added to the set A and u, w to B. See Fig. 8.2 for an example of the construction of $D(I)$.

By construction, note that $D(I)$ is a directed acyclic graph whose underlying graph is bipartite, with bipartition $\{A, B\}$. Furthermore, note that $\lvert V(D)\rvert = n + m + 3$ and, therefore, the construction can be done in linear time, and the solution value, $k + 3$, depends exclusively on the parameter k of the **Set Cover** problem. Therefore, once the equivalence of the instances is proved, the NP-completeness, $O(\log n)$-inapproximability, and W[2]-hardness of **Set Cover** will be inherited to the problem of deciding whether $\overrightarrow{\text{in}}_g(D) \leq p$, given an acyclic orientation of a bipartite graph D and a positive integer p. Note that u is a source and v, w are sinks. Therefore, they are extreme vertices and must belong to every interval set in the geodesic convexity of $D(I)$. In addition, $(u, w) \notin A(D(I))$ and, therefore, (u, f_i, w) is a shortest (u, w)-path in $D(I)$ for each $i \in \{1, \ldots, m\}$.

Let $\mathcal{F}' = \{F_i \mid i \in I\} \subseteq \mathcal{F}$ for some $I \subseteq \{1, \ldots, m\}$ such that $\bigcup_{i \in I} F_i = U$ and $\lvert I\rvert \leq k$. Take $S = \{f_i \mid i \in I\} \cup \{u, v, w\}$. As $\lvert I\rvert \leq k$, note that $\lvert S\rvert \leq k + 3$.

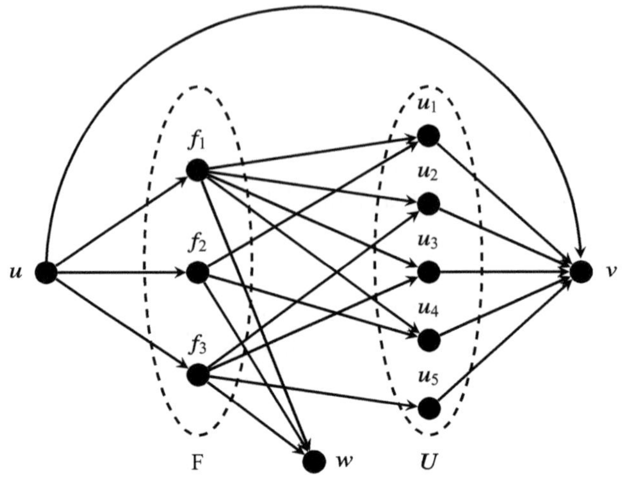

Fig. 8.2 Oriented graph $D(I)$, assuming $U = \{1, 2, 3, 4, 5\}$, $\mathcal{F} = \{F_1 = \{1, 2, 3, 4\}, F_2 = \{1, 4\}, F_3 = \{2, 3, 5\}\}$

As $\mathcal{F}'$ is a cover for U, for each $j \in U$, there is $F_i \in \mathcal{F}$ such that $(f_i, u_j) \in A(D)$, a shortest (f_i, v)-path (f_i, u_j, v) is obtained. Then S is a geodesic interval set of $D(I)$.

On the other hand, let S be a geodesic interval set of $D(I)$ with at most $k + 3$ vertices. As previously argued, we have that $u, v, w \in S$. If there exists $u_j \in S$, observe that we can replace u_j by a vertex f_i such that $(f_i, u_j) \in A(D)$. Thus, we obtain another geodesic interval set S' such that $|S'| \leq k + 3$. In this way, without loss of generality, we assume that $S \setminus \{u, v, w\} \subseteq \{f_i \mid i \in \{1, \dots, m\}\}$. Let $I = \{i \in \{1, \dots, m\} \mid f_i \in S\}$. Note that the family $\mathcal{F}' = \{F_i \in \mathcal{F} \mid i \in I\}$ satisfies $\bigcup_{F_i \in \mathcal{F}'} F_i = U$ and $|\mathcal{F}'| \leq k$. □

As a positive result, Araújo and Arraes (2022) present polynomial algorithms to determine $\overrightarrow{\text{hn}}_g(D)$ and $\overrightarrow{\text{in}}_g(D)$ when D is the orientation of a cactus graph.

Also note that, as mentioned in Sect. 8.1, Haglin and Wolf (1996) show that $\overrightarrow{\text{hn}}_{p3}(T) \leq 2$ for a tournament T, which implies not only that all its convex sets can be obtained in $O(n^4)$, but also that the value of $\text{hn}(T)$ can be computed in constant time, since it is 1 if the tournament is trivial, or 2 otherwise. To conclude, we observe that there is a recent manuscript that brings more contributions on the hull number and the interval number of oriented graphs, not only in the geodesic and $\overrightarrow{P_3}$ convexities, but also in what would be the equivalent to the P_3^* convexity. The interval function $\overrightarrow{I}_{p3*}(u, v)$, in the $\overrightarrow{P_3^*}$ convexity of an oriented graph D, returns, in addition to u and v, all vertices $w \in V(D)$ such that there are arcs $(u, w), (w, v) \in A(D)$, but only if $(u, v) \notin A(D)$ (Araújo et al. 2023). That is, such an interval function returns only vertices in shortest (u, v)-paths of length two in D.

Exercises

Exercise 8.1 Prove Proposition 8.1.

Exercise 8.2 Prove Proposition 8.2.

Exercise 8.3 Prove Propositions 8.5 and 8.6.

Exercise 8.4 Prove Theorem 8.6.

Exercise 8.5 Prove Propositions 8.11and 8.12.

Exercise 8.6 Complement the proof of Theorem 8.13 for the case of split graphs and bipartite graphs.

Chapter 9
Applications of Graph Convexity

9.1 Diffusion Models in Graphs

As seen in Sect. 2.4, Chen (2009) defines a *TSS model* (*target set selection*) as a graph G and a threshold function $\tau : V(G) \to \mathbb{N}$. We assume that $\tau(v) > 0$ for every vertex v; Lemma 2.5 showed that any threshold function of this type induces a *TSS convexity*. Threshold functions can model diffusion processes in graphs and influence propagation in social networks and disease contamination among others.

The diffusion process is as follows: initially the vertices of a given set S_0 are *active*, the remaining vertices are *inactive*, and active vertices remain active forever. At each step, an inactive vertex v becomes active if it has at least $\tau(v)$ active neighbors. The process is synchronous: all inactive vertices update their active/inactive status at the same time at each step of the process. We say that S_t is the set of active vertices at time t and that the *iteration time* $ti(S_0)$ is the smallest value t such that $S_t = S_{t+1}$.

Such processes were investigated under different names: *bootstrap percolation* in Chalupa et al. (1979), *dynamic monopoly* in Peleg (1998), *influence maximization* in Kempe et al. (2003), *target set selection* in Chen (2009), and *irreversible conversion* in Centeno et al. (2011). A set S_0 that activates all vertices at the end of the diffusion process is called a *target set* in Chen (2009), which basically consists of a *hull set* in the corresponding convexity.

Historically, this research line focused on *majority thresholds* (Peleg 1998), with the majority threshold function $\tau(v) = \lceil d(v)/2 \rceil$, where $d(v)$ is the degree of v, and also on the constant value threshold function r, called *r-neighbor bootstrap percolation* (Chalupa et al. 1979). In addition, it focused on probabilistic results, when thresholds are randomly chosen within a given range (Kempe et al. 2003) or when the initial set S_0 is randomly chosen (Holroyd 2003). As already mentioned,

J. Araújo et al., *Introduction to Graph Convexity*, SBMAC Collection on Applied and Computational Mathematics, https://doi.org/10.1007/978-3-031-84128-6_9

the most general definition for any threshold function was given by Chen (2009) in the context of *target set selection* (TSS).

The most investigated problem in this area is the TSS-SIZE problem: determining the size of the smallest target set, i.e., calculating the hull number $\mathrm{hn}(G)$ in the corresponding convexity. See, for example, Ben-Zwi et al. (2011), Nichterlein et al. (2013), Chopin et al. (2014), Bazgan et al. (2014), Ehard and Rautenbach (2019).

Besides the TSS-SIZE, other problems have been recently investigated, such as TSS-TIME, TSS-MAX-CONVEX, and TSS-DOMINATION, which basically consist in computing the percolation time $\mathrm{tp}(G)$, the convexity number $\mathrm{con}(G)$, and the interval number $\mathrm{in}(G)$ of the graph in the corresponding convexity.

The TSS-TIME problem was first studied by Flocchini et al. (2003). It was studied with constant thresholds $\tau(v) = 2$ by Marcilon and Sampaio (2018b) and with any thresholds by Keiler et al. (2023), who also obtained a polynomial algorithm for trees and proved NP–hardness for bipartite graphs. A motivation for this problem is to determine the longest time to wait for the entire network to become active.

The problem TSS-MAX-CONVEX was first studied by Araújo and Sampaio (2023), who also proved that it is Poly-APX–complete even in split graphs and in bipartite graphs and obtained linear algorithms for cographs and distance–hereditary graphs, when the maximum threshold is a constant. A motivation for this problem is that it is a natural measure for the size of *closed communities*, which remain inactive even if all other vertices are activated.

The problem TSS-DOMINATION was investigated under different names: VECTOR DOMINATION by Cicalese et al. (2013), $(1, |V(G)|)$-TARGET SET SELECTION by Cicalese et al. (2014), and MIN-TBIDS by Eirinaki et al. (2016). Polynomial algorithms were obtained for trees and cographs by Cicalese et al. (2013), graphs with bounded *treewidth* by Cicalese et al. (2014), and split-indifferent graphs by Mafort and Protti (2020).

To illustrate some results, we show how to solve in polynomial time for trees the problems TSS-SIZE and TSS-TIME, which want to find target sets with minimum size and maximum percolation time, respectively. Vertices with $\tau(v) > d(v)$ must be in any target set and, thus, we can simulate the diffusion process from these vertices until it stops. For simplicity, we assume that $0 < \tau(v) \leq d(v)$ for every vertex v.

TSS-SIZE in trees is solved by the algorithm below from Chen (2009). Figure 9.1 shows an example of the execution of this algorithm. The numbers inside the vertices represent the order in which the vertices were selected by the algorithm to decide their status inside/outside the target set. Note that the percolation time of this target set is 5, much lower than the maximum 15 of Fig. 9.2.

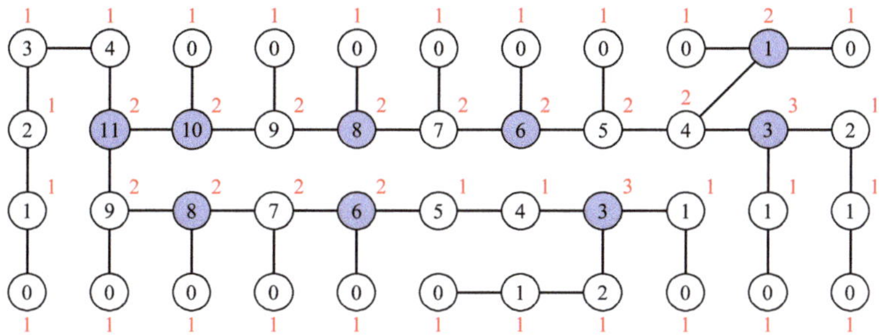

Fig. 9.1 Tree with minimum target set in blue: 9 vertices. Thresholds in red. The number in the vertices is the order in which the status white/blue has changed. The vertex 11 is the root chosen by the algorithm

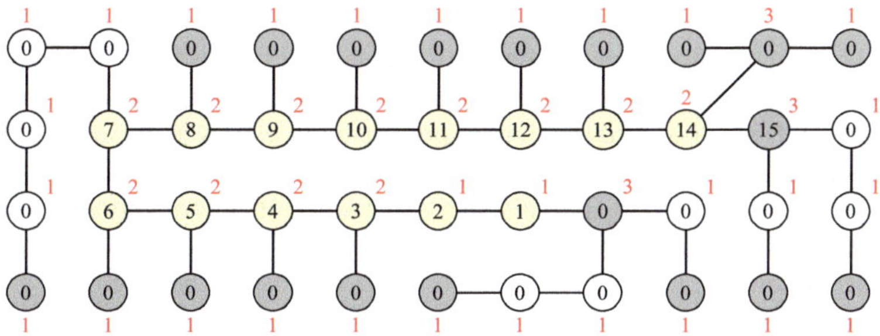

Fig. 9.2 Tree with percolation time 15. Thresholds in red, saturated vertices in gray, and maximum non-saturated path in yellow. The number in the vertices are activation times. Time 0 represents the target set, with 29 vertices

- Algorithm TSS-SIZE-TREE (tree T, threshold function τ)

consider T as a tree rooted at a vertex r of degree $d(r) \geq 2$
let $\tau'(v) = \tau(v)$ for every vertex v of T
let $x(f) = 0$ for every leaf f of T
while there is a vertex v with $x(v)$ not defined, *do*
    let u be a vertex whose children have $x(\cdot)$ defined and *let* w be the parent of u
    if $\tau'(u) \geq 2$, *then*
        $x(u) \leftarrow 1$; $\tau'(w) \leftarrow \tau'(w) - 1$
    else
        $x(u) \leftarrow 0$
        if $\tau'(u) \leq 0$, *then:* $\tau'(w) \leftarrow \tau'(w) - 1$
return the vertices v with $x(v) = 1$

TSS-TIME in trees is solved by Keiler et al. (2023): the maximum percolation time is associated with a longest path in the tree whose internal vertices are *not saturated*, where a vertex is *saturated* if the threshold is greater or equal to the degree. Figure 9.2 shows an example for the same tree of Fig. 9.1, with the path of non-saturated vertices in yellow, where the numbers inside the vertices represent the time at which the vertex was activated by the target set, represented by the vertices with time 0. Note that the maximum time 15 is obtained by a target set with 29 vertices, much higher than the minimum 9 from Fig. 9.1.

As open questions, we have the problems of computating the other convexity parameters related to the TSS models such as, for example, TSS-CARATHÉODORY, TSS-RADON, TSS-HELLY, and TSS-RANK.

9.2 Graph Convexity Games

Interval Game and Closed Interval Game

As stated before, the first paper on convexity in general graphs published in English is the paper "Convexity in graphs", by Harary and Nieminem (1981). Three years later, Harary (1984) proposed the first graph convexity games in the abstract Harary (1984). Below we describe the oldest game: the *geodesic interval game* studied by Buckley and Harary (1985b). Two players, Alice and Bob, alternately select vertices not yet selected, starting with Alice. Let S be the set of vertices selected during the game, initially empty. Each player adds to S one (and only one) vertex $v \notin S$. The games end when $I_g(S) = V(G)$.

This game has two variants: the *normal game* (the last to play wins) and the *misère game*[1] (the last to play loses). The normal and misère variants are also called *achievement and avoidance game*, respectively. From the classic theorem of Zermelo (1913), one of the two players has a winning strategy in each of these games, as they are finite games without a draw and with perfect information. Thus, the objective of each of the games is to decide whether Alice has a winning strategy or not. To illustrate, we have the simple lemma below, whose proof was left for Exercise 9.5.

Lemma 9.1 (Buckley and Harary 1985b) *Consider the geodesic interval game on the cycle C_n. Alice wins the normal variant if and only if n is odd and wins the misère variant if and only if $n \mod 4$ is 1 or 2.*

Some results of Buckley and Harary (1985b) were improved by Nečásková (1988) and, years later, Haynes et al. (2003) obtained results for trees and complete multipartite graphs in the geodesic interval game.

The *closed interval game* is a similar game defined by Buckley and Harary (1985a). Let S be the set of vertices selected during the game, initially empty. Alice and Bob alternately add to S one vertex $v \notin I_g(S)$. The games end when $I_g(S) = V(G)$. Recall that the interval game allows the selection of a vertex in

[1] This name is because the players intentionally play with the *aim of losing* as, for example, a checkers player trying to force the opponent to take all his pieces.

$I_g(S) \setminus S$, unlike the closed interval game. Only recently, the normal and misère variants of the closed geodesic interval game were solved for trees (Araújo et al. 2024a) and for cacti and block graphs (Dailly et al. 2024), using the Sprague–Grundy theory. The partizan variants of this game were also solved recently in trees by Araújo et al. (2024b) using surreal numbers and the combinatorial game theory. Similar games related to the construction of a convex set in a graph were studied recently by Benesh et al. (2024) and by Brosse et al. (2025).

Considering the interval game and the closed interval game as the problems of deciding whether Alice has a winning strategy, it is not difficult to show that these games are PSPACE. However, the PSPACE–hardness of these games (normal and misère variants) is still an open question.

General Position Game

In the interval games, the final set S of selected vertices is an interval set ($I_g(S) = V(G)$), but it may not be minimal, i.e., S may contain a proper subset that is also an interval set. For example, in the path P_n $v_1 \dots v_n$, a possible (although unlikely) sequence of choices would be $v_1, \dots, v_n$ with Alice (resp. Bob) selecting the vertices v_i with i odd (resp. even). In this example, all vertices could be selected, although there is only a minimal interval set $S = \{v_1, v_n\}$, which is also minimum.

To avoid this situation, Klavžar et al. (2022) recently introduced the *geodesic general position game*, with the normal and misère variants. In this game, the set S of selected vertices must always be in geodesic general position, i.e., there cannot be a vertex $v \in S$ in $I_g(S \setminus \{v\})$. With this, the game ends when it is no longer possible to choose vertices and, therefore, in the end S will not necessarily be an interval set as in the other games. It is easy to confuse this game with the closed geodesic interval game, but they are quite different. In the same example of the path P_n $v_1 \dots v_n$ for $n \geq 3$, Alice chooses any vertex v_i, Bob chooses a vertex v_j, and then Alice cannot choose the third vertex of S, losing the normal game and winning the misère general position game.

The lemma below solves the normal game in bipartite graphs. The problem of the misère general position game in bipartite graphs remains open.

Lemma 9.2 *Given a bipartite graph G, Alice wins the geodesic general position normal game if and only if the number of isolated vertices of G is odd.*

Proof Exercise 9.6 (Hint: Lemma 3.4). □

Regarding computational complexity, Chandran et al. (2024) proved that the normal and misère variants of the geodesic general position game are PSPACE–hard. To exemplify a proof of PSPACE–hardness, we show the simplest proof.

Theorem 9.1 (Chandran et al. 2024) *The normal variant of the geodesic general position game is* PSPACE*–complete in graphs with diameter 4.*

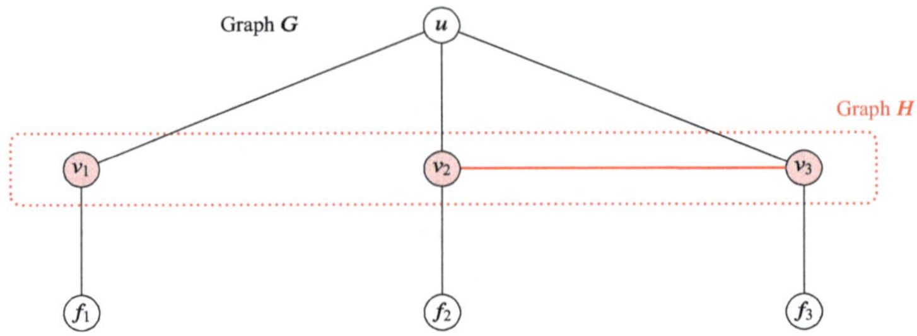

Fig. 9.3 Alice wins the normal variant of the general position game in G if and only if player 2 wins the clique forming game in H

Proof We obtain a reduction from the clique forming game, proved PSPACE–complete by Schaefer (1978). In this game, players 1 and 2 alternately select vertices of a graph G and the set of chosen vertices must be a clique. The first unable to play loses. It is strongly related to the classic game *Node Kayles*,[2] in which one wants to obtain an independent set instead of a clique. The clique forming game is the *Node Kayles* game played on the complement of the graph and vice versa.

It is not difficult to prove that the normal variant of the general position game is PSPACE. Let H a clique forming game instance with $V(H) = \{v_1, \ldots, v_n\}$. We obtain a graph G such that Alice has a winning strategy in the general position game in G if and only if player 2 of the clique forming game in H has a winning strategy.

Let G be the graph obtained from H by adding a new vertex u adjacent to all vertices of H and adding a new vertex f_i, for each vertex v_i of H, whose only neighbor is v_i. Note that G has diameter 4 (see Fig. 9.3).

If Alice chooses v_i first, Bob wins by choosing f_i, as the geodesic between f_i and any other vertex of G passes through v_i. Analogously if Alice selects f_i. We can then assume that Alice chooses u in the first move.

Note that from then on Alice and Bob cannot choose nonadjacent vertices v_i and v_j, as $v_i - u - v_j$ is a geodesic. In addition, they cannot choose f_i and f_j such that v_i and v_j are nonadjacent, as $f_i - v_i - u - v_j - f_j$ is a geodesic. They also cannot choose v_i and f_j such that v_i and v_j are nonadjacent, as $v_i - u - v_j - f_j$ is a geodesic. Finally, they cannot choose v_i and f_i, as $u - v_i - f_i$ is a geodesic.

Let $C = \{v_i \mid v_i \text{ or } f_i \text{ was chosen}\}$. From the previous paragraph, we have that C is a clique of H and we can assume that the players do not choose vertices $f_1, \ldots, f_n$, as selecting f_i is essentially the same as selecting v_i.

Therefore, if player 2 of the clique forming game in H has a winning strategy, Alice has a winning strategy in the normal variant of the general position game in

[2] The name *Kayles* is an English version of the French term *Quilles*, which refers to an old European variant of the bowling game, played on grass. The mathematical game *Kayles* was introduced by Dudeney (1908). The game *Node Kayles* is the bowling version in graphs, in which one selects in each move a vertex, which is *knocked down* along with its neighbors, obtaining in the end a maximal independent set from the selected vertices.

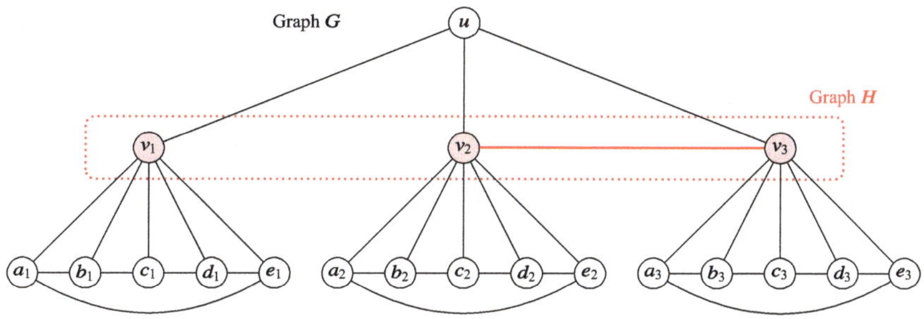

Fig. 9.4 Alice wins the misère variant of the general position game in G if and only if player 2 wins the clique forming game in H

G, as she will be the second to play in H (remember that u must be the first vertex to be chosen). In addition, if player 1 of the clique forming game in H has a winning strategy, Bob also has in G, as he will be the first to play on the vertices of H. □

Theorem 9.2 (Chandran et al. 2024) *The misère variant of the geodesic general position game is* PSPACE*–complete in graphs with diameter 4.*

Proof Exercise 9.7. Strongly based on the reduction of Theorem 9.1, with the small modification described in Fig. 9.4. □

Finally, we can also define the *geodesic convex position game*, with normal and misère variants, the difference being that the selected vertices must not only be in general position, but also in convex position. This game was proved PSPACE-complete by Araújo et al. (2024a) in both normal and misère variants.

Hull Game and the Optimization Variants

Harary (1984) mentioned a game "*involving the convex hull*," but gave no formal definition. Araújo et al. (2024a) introduced a natural game of graph convexity, related to the hull of the set S of selected vertices, called *hull game*. The game is natural in the sense that it simulates an activation process in the graph, which ends when every vertex is activated. That is, Alice and Bob select vertices alternately until $\text{conv}_C(S) = V(G)$, where S is the set of vertices selected so far.

Since there is the closed interval game (Buckley and Harary 1985a), we can also define the *closed hull game*, the only difference to the hull game being that they cannot select vertices of $\text{conv}_C(S)$, the convex hull of the vertices already selected.

Similarly, the interval, general position, and convex position games are defined for any graph convexity C. For example, with C being the P_3 convexity, we have the P_3 (closed) interval game, the P_3 (closed) hull game, the P_3 general position game, and the P_3 convex position game.

In addition to the normal and misère variants of these games, defined earlier, there is also the *optimal variant* (or *optimization variant*), quite common in other game problems, such as in the *graph coloring game* (Costa et al. 2020). In the optimal variant of the hull game, the instance also has an integer k, in addition to the graph G, and the goal is to decide if Alice has a winning strategy in which the set S of selected vertices ends with at most k vertices. That is, Alice is the cooperative player who wants to optimize S (by minimizing it), while Bob is the noncooperative player who wants to annoy Alice (by maximizing S), no matter who ends the game (unlike the other variants).

The optimal variants of the other games seen in this chapter can also be defined. In the optimal variant of the interval game, the goal is to decide if Alice has a winning strategy in which S ends with at most k vertices. In the optimal variant of the general position game and the convex position game, the goal is to decide if Alice has a winning strategy, in which S ends with at least k vertices. Note the change from *at most* to *at least*, since naturally one wants to obtain a minimum hull set or interval set and a maximum general position set or convex position set.

Optimal variants of games usually define game parameters such as, for example, the chromatic game number (Costa et al. 2020). With this, convexity game parameters of the optimization variants of the six games are defined:

- *Game hull number* $\mathrm{ghn}_C(G)$
- *Game interval number* $\mathrm{gin}_C(G)$
- *Closed game hull number* $\mathrm{cghn}_C(G)$
- *Closed game interval number* $\mathrm{cgin}_C(G)$
- *Game general position number* $\mathrm{ggp}_C(G)$
- *Game rank* $\mathrm{grk}_C(G)$

which are basically the optimal values of k for which Alice has a winning strategy. In the case of the interval and the hull games, the minimum k is desired, and in the case of the general position and the convex position games, the maximum k is desired. The lemma below shows simple inequalities between the parameters of convexity games.

Lemma 9.3 (Araújo et al. 2024a) *Given a convexity C on a graph G:*

- $\mathrm{hn}_C(G) \leq \mathrm{cghn}_C(G) \leq \mathrm{ghn}_C(G) \leq \min\left\{ 2 \cdot \mathrm{hn}_C(G) - 1,\ n \right\}$.
- $\mathrm{in}_C(G) \leq \mathrm{gin}_C(G) \leq \min\left\{ 2 \cdot \mathrm{in}_C(G) - 1,\ n \right\}$.
- $\mathrm{in}_C(G) \leq \mathrm{cgin}_C(G) \leq n$.
- $\mathrm{gp}_C(G) \geq \mathrm{ggp}_C(G)$ *and* $\mathrm{rk}_C(G) \geq \mathrm{grk}_C(G)$ *and* $\mathrm{ggp}_C(G) \geq \mathrm{grk}_C(G)$.

Proof Exercise 9.8. Note that the restrictions of closed games only benefit Alice and she can always play on the vertices of a minimum hull or interval set. Furthermore, every convex position is a general position. □

The lemma below obtains a result for all games seen in this chapter on the complete graph K_n in the P_3, geodesic, and monophonic convexities.

Lemma 9.4 (Araújo et al. 2024a) *Let $n \geq 2$. In the P_3 convexity, Alice loses the normal variant and wins the misère variant of all games in this chapter on K_n. In the geodesic and monophonic convexities, Alice wins all games on K_n if and only if n is odd in the normal variant or n is even in the misère variant. Furthermore, all game parameters on K_n are equal to n in the monophonic and geodesic convexities and are equal to 2 in the P_3 convexity.*

Proof In the P_3 convexity, all games end on the second move. In the monophonic and geodesic convexities, every vertex must be selected in all games. □

The lemma below obtains a result for all games seen in this chapter on the cycle C_n in the monophonic convexity.

Lemma 9.5 (Araújo et al. 2024a) *Let $n \geq 4$. Bob wins the normal and misère variants of all games in this chapter in the monophonic convexity on the cycle C_n, except for the misère game of the general position and the convex position, which are always won by Alice. Furthermore, every game parameter is equal to 3 in C_n in the monophonic convexity, except for* $\text{ggp}_\text{m}(C_n) = \text{grk}_\text{m}(C_n) = 2$.

Proof In the normal game, Bob chooses a vertex not adjacent to the first vertex chosen by Alice. In the misère game, Bob chooses a neighbor. □

Finally, it is not difficult to see that, despite the graphs constructed in the proofs of Theorems 9.1 and 9.2 are not necessarily distance–hereditary, the arguments serve both for geodesic and monophonic convexity. In addition, they are not only valid for general position, but also for convex position. With this, we have the following:

Corollary 9.1 *The general position and convex position games are* PSPACE–*complete in the normal and misère variants in the geodesic and monophonic convexities, even in graphs with diameter 4.*

Hull Game and Convex Geometries

In this subsection, we analyze the hull game, defined in the previous section, a little more closely and present an interesting relationship between convex geometries (Chap. 4) and winning strategies in the hull game.

We initially show some examples reaching the upper and/or lower bound of the first inequality of Lemma 9.3 with respect to the P_3 convexity. An example at the lower bound of Lemma 9.3 is K_n (see Lemma 9.4). For the upper bound, we have the cycles C_4 and C_6, since $\text{hn}_{\text{p3}}(C_4) = 2$, $\text{cghn}_{\text{p3}}(C_4) = \text{ghn}_{\text{p3}}(C_4) = 3$, $\text{hn}_{\text{p3}}(C_6) = 3$, $\text{cghn}_{\text{p3}}(C_6) = 4$, and $\text{ghn}_{\text{p3}}(C_6) = 5$. For the lower bound, we have the cycle C_5, as $\text{hn}_{\text{p3}}(C_5) = \text{cghn}_{\text{p3}}(C_5) = \text{ghn}_{\text{p3}}(C_5) = 3$. An example in the middle is the cycle C_7, as $\text{hn}_{\text{p3}}(C_7) = 4$ and $\text{cghn}_{\text{p3}}(C_7) = \text{ghn}_{\text{p3}}(C_7) = 5 < 7$.

Regarding closed games in the geodesic convexity on cycles, Araújo et al. (2024a) proved the following lemma.

Lemma 9.6 (Araújo et al. 2024a) *Let $n \geq 4$ and consider the geodesic convexity. Alice wins the normal variant of the closed hull game and closed interval game on C_n if and only if n is odd. Furthermore, Bob always wins the misère variant of the closed hull game and closed interval game on C_n. Finally,* $\mathrm{cghn_g}(C_n) = \mathrm{cgin_g}(C_n) = \mathrm{ghn_g}(C_n) = \mathrm{gin_g}(C_n) = 3$.

Proof Consider C_n as the cycle $v_1, v_2, \ldots, v_n$ and assume that Alice selects v_1 in her first move. In the misère variant, Bob always selects the vertex $v_{\lfloor n/2 \rfloor}$ and Alice loses on the next move. In the normal variant, if n is even, Bob selects $v_{\lceil n/2 \rceil}$ and wins. Otherwise, Alice manages to win on the next move. □

In addition to these results for simple graphs, Araújo et al. (2024a) also obtain an interesting connection between convex geometries in graphs, seen in Chap. 4, and winning strategies in hull games.

Theorem 9.3 (Araújo et al. 2024a) *Let C be a convex geometry[3] on G. Alice wins the hull game on G in convexity C if and only if n is odd in the normal variant or n is even in the misère variant. Furthermore,*

$$\mathrm{ghn}_C(G) = \min\left\{\, 2 \cdot |\,\mathrm{Ext}_C(G)| \; - 1, \; n \right\}.$$

Proof Let $F = \mathrm{Ext}_C(G)$. As C is a convex geometry on G, then F is a hull set. As $V(G)\setminus\{f\}$ is convex for every $f \in F$, every vertex of F must be selected during the interval game and the hull game. Therefore the hull game ends when the last vertex of F is selected. In the normal variant, if n is odd, Alice plays avoiding the last two vertices of F forcing Bob to select the penultimate vertex of F, causing Alice to win on the next move. If n is even, Bob wins following this same argument. In the misère variant, if n is even, Alice plays avoiding the last vertex of F, forcing Bob to select it and lose the game. If n is odd, Bob wins following this same argument. □

The monophonic, geodesic, and P_3 convexities are treated below.

Corollary 9.2 (Araújo et al. 2024a) *In Ptolemaic graphs (resp. chordal), Alice wins the geodesic (resp. monophonic) hull game if and only if n is odd in the normal variant or n is even in the misère variant.*

Corollary 9.3 (Araújo et al. 2024a) *Let T be a rooted tree, in which every non-leaf vertex has at least two children. Then Alice wins the P_3 hull game on T if and only if n is odd in the normal variant or n is even in the misère variant. Furthermore,* $\mathrm{ghn_{p3}}(T) = n$.

Proof Note that $\mathrm{Ext_{p3}}(T)$ is the set of leaves of T, as well as being also a P_3 hull set of T, since every internal vertex has two children. Therefore, following

[3] See Chap. 4.

the arguments in the proof of Theorem 9.3, we have the result. Finally, $n \leq 2 \cdot |\,\mathrm{Ext}_{\mathrm{p3}}(T)| - 1$. □

Finally, Araújo et al. (2024a) also obtain PSPACE–completeness results for hull games in two important graph convexities.

Theorem 9.4 (Araújo et al. 2024a) *The hull and closed hull games in the geodesic and monophonic convexities in the normal and misère variants are* PSPACE–*complete.*

Exercises

Exercise 9.1 Prove that $\mathrm{hn}_{\mathrm{p3}}(C_4) = 2$ and $\mathrm{cghn}_{\mathrm{p3}}(C_4) = \mathrm{ghn}_{\mathrm{p3}}(C_4) = 3$.

Exercise 9.2 Prove that $\mathrm{hn}_{\mathrm{p3}}(C_5) = \mathrm{cghn}_{\mathrm{p3}}(C_5) = \mathrm{ghn}_{\mathrm{p3}}(C_5) = 3$.

Exercise 9.3 Prove that $\mathrm{hn}_{\mathrm{p3}}(C_6) = 3$, $\mathrm{cghn}_{\mathrm{p3}}(C_6) = 4$ and $\mathrm{ghn}_{\mathrm{p3}}(C_6) = 5$.

Exercise 9.4 Prove that $\mathrm{hn}_{\mathrm{p3}}(C_7) = 4$ and $\mathrm{cghn}_{\mathrm{p3}}(C_7) = \mathrm{ghn}_{\mathrm{p3}}(C_7) = 5$.

Exercise 9.5 Prove Lemma 9.1.

Exercise 9.6 Prove Lemma 9.2 (Hint: Lemma 3.4).

Exercise 9.7 Prove Theorem 9.2 (Hint: Theorem 9.1 and Fig. 9.4).

Exercise 9.8 Prove Lemma 9.3.

Appendix A
Graph Theory

A (simple) graph G is formed by a set of *vertices*, denoted by $V(G)$, and a set of *edges*, denoted by $E(G)$. Each edge is an (unordered) pair of distinct vertices. If xy is an edge, then the vertices x and y are the *ends* of this edge. We also say that x and y are *connected*, are *adjacent*, or are *neighbors*. A graph can be represented geometrically as a set of points in the plane (representing the vertices) and lines connecting these points (representing the edges). We note that the same graph can have several different geometric representations.

Example A.1 Let G be the graph such that $V(G) = \{a, u, v, w, x, y, z\}$ and $E(G) = \{uv, vw, wx, xy, yz, zu, av, ax, az\}$. In Fig. A.1, we have two different geometric representations for G.

We use the notation $n = |V(G)|$ and $m = |E(G)|$. The *order* of G is n and its *size* is $n + m$. The *trivial graph* has only one vertex. The *null graph* has $V(G) = \emptyset$.

A *multigraph* generalizes the concept of a simple graph. In a multigraph, there may exist *parallel* or *multiple edges* (edges with the same ends) and *loops* (edges of the form xx). *Digraphs* (or *directed graphs*) also generalize simple graphs. In a digraph, the edges are ordered pairs of distinct vertices, often called *arcs*. That is, xy and yx represent distinct edges (arcs).

A.1 Neighborhood, Degree, Subgraphs, and Complement

The *neighborhood* of a vertex v is the set of its neighbors. We use the notation $N(v)$ to stand for the neighborhood of v. The *closed neighborhood* of a vertex v is defined as $N[v] = N(v) \cup \{v\}$. The *degree* of a vertex is the number of times it occurs as the endpoint of an edge. (This definition applies to both graphs and multigraphs.) We use the notation $d(v)$ to designate the degree of vertex v. In a simple graph, the

J. Araújo et al., *Introduction to Graph Convexity*, SBMAC Collection on Applied and Computational Mathematics, https://doi.org/10.1007/978-3-031-84128-6

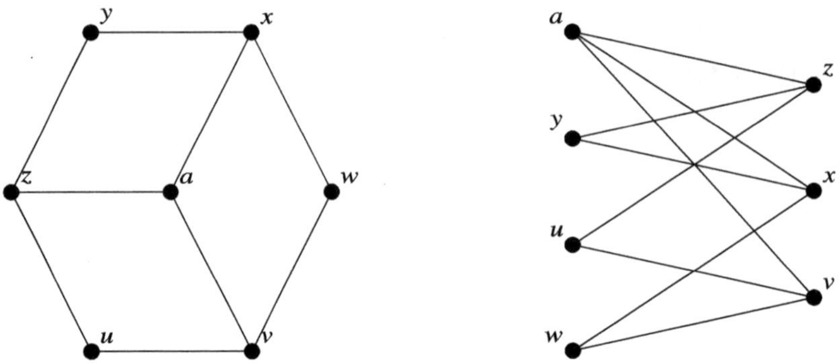

Fig. A.1 Two different geometric representations for the same graph

degree of a vertex is equal to the number of neighbors it has, that is, $d(v) = |N(v)|$.

A graph is *regular* when all its vertices have the same degree. A graph is *k-regular* when all its vertices have a degree equal to k.

The *maximum degree* of G is defined as $\Delta(G) = \max\{d(v) \mid v \in V(G)\}$. The *minimum degree* of G is defined as $\delta(G) = \min\{d(v) \mid v \in V(G)\}$.

Given a graph G such that $V(G) = \{v_1, v_2, \ldots, v_{n-1}, v_n\}$ and the degrees of the vertices satisfy $d(v_1) \leq d(v_2) \leq \cdots \leq d(v_{n-1}) \leq d(v_n)$, the *degree sequence* of G is precisely the sequence $(d(v_1), d(v_2), \ldots, d(v_{n-1}), d(v_n))$.

Example A.2 The degree sequence of the graph G defined earlier in Example A.1 is $(2, 2, 2, 3, 3, 3, 3)$. We have that $\delta(G) = 2$ and $\Delta(G) = 3$.

A vertex is *isolated* when it has degree zero (it has no neighbors). A vertex v is *universal* when it is connected by edges to all other vertices, that is, $N(v) = V(G) \setminus \{v\}$. If v is a universal vertex, then $d(v) = n - 1$.

The following theorem is known as the *handshaking theorem*:

Theorem A.1 *In any simple graph G, $\sum_{v \in V(G)} d(v) = 2m$.*

Proof Note that each edge xy is counted twice in the sum $\sum_{v \in V(G)} d(v)$ – once in the term $d(x)$ and another in the term $d(y)$. □

A *subgraph* of a graph G is a graph H such that $V(H) \subseteq V(G)$ and $E(H) \subseteq E(G)$. H is a *proper subgraph* of G when H is a subgraph of G that is not G itself. A *spanning subgraph* of G is a subgraph H of G such that $V(H) = V(G)$. In other words, H has the same vertices as G, but not necessarily all the edges of G. A subgraph H of G is an *induced subgraph by a set of vertices* $X \subseteq V(G)$ if $V(H) = X$ and H has the following property: if $xy \in E(G)$ and $x, y \in X$, then $xy \in E(H)$. In this case, we use the notation $H = G[X]$. Informally, an induced subgraph by a set of vertices X maintains all the original edges of G that have both ends in X. A subgraph H of G is an *induced subgraph by a set of edges* $E' \subseteq E(G)$

if (a) $E(H) = E'$; (b) $V(H) = \{x \mid x$ is an end of some edge of $E'\}$. We use the notation $H = G[E']$ to designate that H is a subgraph induced by a set of edges E'.

The following notation is quite useful. If S is a subset of vertices of G, then $G - S = G[V(G) \setminus S]$. If v is a vertex of G, then $G - v = G - \{v\}$. If E' is a subset of edges of G, then the graph $G - E'$ is defined as follows: $V(G - E') = V(G)$ and $E(G - E') = E(G) \setminus E'$. If e is an edge of G, then $G - e = G - \{e\}$.

The *union* $G \cup H$ of two graphs G and H is the graph with $V(G \cup H) = V(G) \cup V(H)$ and $E(G \cup H) = E(G) \cup E(H)$. The *intersection* $G \cap H$ of two graphs G and H is the graph with $V(G \cap H) = V(G) \cap V(H)$ and $E(G \cap H) = E(G) \cap E(H)$.

Two graphs G and H are *disjoint in vertices* if $V(G) \cap V(H) = \emptyset$. Two graphs G and H are *disjoint in edges* if $E(G) \cap E(H) = \emptyset$. If G and H are disjoint in vertices, then it is clear that they are also disjoint in edges. However, G and H can be disjoint in edges having some vertices in common.

The *complement* of a graph G is the graph $\overline{G}$ such that $V(\overline{G}) = V(G)$ and $E(\overline{G}) = \{xy \mid xy \notin E(G)\}$. Note that G and $\overline{G}$ are graphs disjoint in edges. Therefore, $G \cap \overline{G}$ is a graph without edges. In addition, $G \cup \overline{G}$ is a complete graph.

Example A.3 If G is the graph of Example A.1, then $\overline{G}$ is the graph of Fig. A.2.

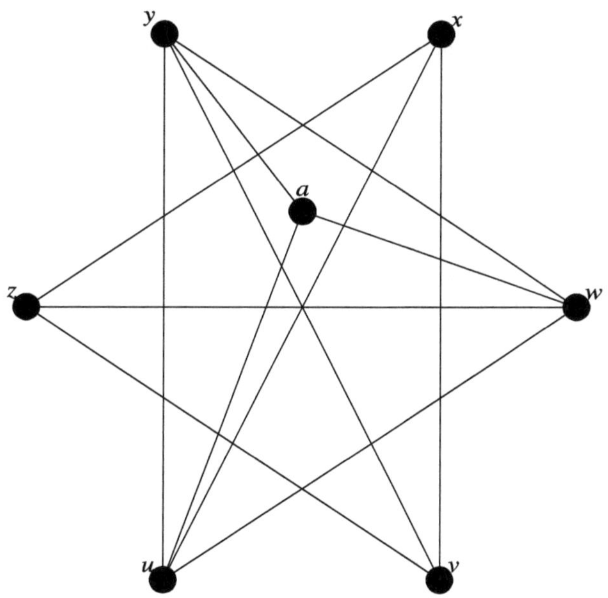

Fig. A.2 Geometric representation of $\overline{G}$, where G is the graph of Exercise A.1

A.2 Cliques, Paths, and Cycles

A graph G is *complete* if any two vertices are neighbors. We denote by K_n the complete graph with n vertices: K_1 is the trivial graph, K_2 has two vertices and one edge, and K_3 is the triangle. Figure A.3 shows the graphs K_3, K_4, and K_5.

A *clique* in G is a subset $K \subseteq V(G)$ such that $G[K]$ is complete (any two vertices are adjacent). The cardinality of a largest clique in G is denoted by $\omega(G)$. An *independent set* in G is a subset $S \subseteq V(G)$ such that $G[S]$ is a graph without edges (any two vertices are nonadjacent). The cardinality of a largest independent set in G is denoted by $\alpha(G)$.

A *walk* is a sequence of vertices $v_1 v_2 \dots v_k$ such that $v_j v_{j+1} \in E(G)$ for $1 \leq j < k$. There may be repetition of vertices and edges in a walk. If $v_1 = v_k$, we say the walk is *closed*; otherwise, it is *open*. A *trail* is a walk without repetition of edges, but there may be repetition of vertices. A *path* is a trail without repetition of vertices. The *length* of a path is its number of edges. If P is a path and u, v are vertices of this path, we denote by $P[u, v]$ the subpath of P that goes from u to v. A *cycle* is a closed trail $v_1 v_2 \dots v_{k-1} v_k$ with $k \geq 4$ such that $v_1 v_2 \dots v_{k-1}$ is a path. The length of a cycle is its number of edges (or vertices).

A *chord* is an edge that connects two nonconsecutive vertices of a cycle (or path). An *induced cycle* (resp. *path*) in G is a cycle (resp. path) without chords. We use the notations P_n and C_n for the path and the induced cycle with n vertices, respectively.

Example A.4 Consider again the graph G from Example A.1. Then: $uvazyxaz$ is an open walk, $uvazyxazu$ is a closed walk, $avwxazy$ is an open trail, $uvwxazy$ is a path, $uvwxy$ is an induced path, $uvwxyzu$ is a cycle, and $uvazu$ is an induced cycle.

Remark A.1 Often, walks, trails, paths, and cycles are considered as graphs (or subgraphs) rather than sequences of vertices. For example, we can refer to a path P with k vertices as a graph P such that $V(P) = \{v_1, \dots, v_k\}$ and $E(P) = \{v_j v_{j+1} \mid 1 \leq j < k\}$.

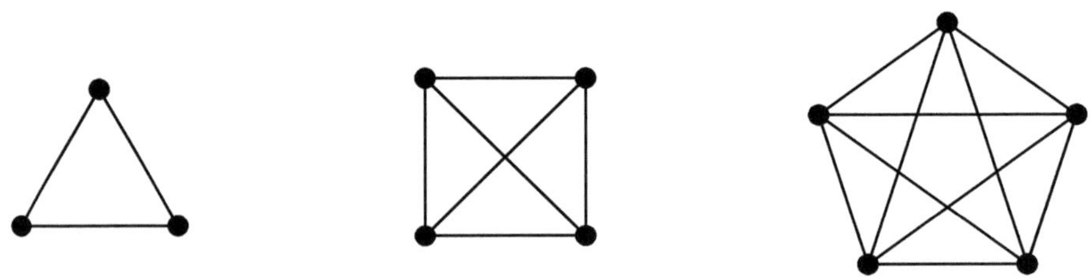

Fig. A.3 From left to right: graphs K_3, K_4, and K_5

A.3 Distance, Diameter, Optimality, and Isomorphism

The *distance* between vertices x and y is the length of the shortest path from x to y in the graph. We use the notation $dist(x, y)$ for the distance between x and y. For every vertex x, $dist(x, x) = 0$. The *eccentricity* of a vertex v in a graph G is defined as $exc(v) = \max\{dist(v, x) \mid x \in V(G)\}$. The *diameter* of a graph G is defined as $diam(G) = \max\{exc(v) \mid v \in V(G)\}$. The *center* of a graph G is the set of vertices of G with minimum eccentricity.

Two graphs G and H are *isomorphic* if there exists a bijection $f : V(G) \to V(H)$ such that $xy \in E(G)$ if and only if $f(x)f(y) \in E(H)$. Informally G and H are the *same* graph, except for different labels for the vertices. We use the notation $G \cong H$ to denote that G and H are isomorphic. Let G and H be any graphs. If there exists some subgraph G' of G that is isomorphic to H, we say that G *contains* H. If there exists some induced subgraph G' of G that is isomorphic to H, we say that G *contains* H *as an induced subgraph*. If no induced subgraph of G is isomorphic to H, we say that G is *free of* H.

A set S is *minimal* (resp. *maximal*) with respect to a property P if (a) S satisfies P and (b) there is no set $S' \subsetneq S$ (resp. $S' \supsetneq S$) that satisfies P. A set S is *minimum* (resp. *maximum*) with respect to a property P if (a) S satisfies P and (b) there is no set S' that satisfies P with $|S'| < |S|$ (resp. $|S'| > |S|$). Every minimum (resp. maximum) set is also minimal (resp. maximal), but not every minimal (resp. maximal) set is minimum (resp. maximum). The concepts *minimal/minimum* and *maximal/maximum* also apply to graphs and subgraphs.

Example A.5 In the graph of Example A.1, $S_1 = \{u, w, y\}$ and $S_2 = \{a, v, x, z\}$ are maximal independent sets, but only S_2 is maximum.

Example A.6 A subset $C \subseteq V(G)$ is a *cover (by vertices)* of G if every edge has at least one endpoint in C. With S_1 and S_2 being the same as in the previous example, S_1 and S_2 are minimal covers of G, but only S_1 is minimum.

A.4 Connected Graphs, Trees, and Bipartite Graphs

A graph G is *connected* if there is a path between any pair of vertices of G. Otherwise, G is *disconnected.* A *connected component* of a graph G is a maximal connected subgraph of G. We denote by $w(G)$ the number of connected components of G. It is clear that G is connected if, and only if, $w(G) = 1$.

A *tree* T is a connected graph without cycles. A *forest* is a graph whose connected components are trees. The vertices of a tree are often called *nodes*, the vertices of degree 1 are the *leaves*, and the others are the *internal nodes*. It is easy to prove that every tree has $m = n - 1$ and at least 2 leaves. We say that a tree is *rooted* when a special vertex r called *root* with degree greater than 1 is defined on it. In a rooted tree, the first vertex on the path from a vertex v to the root is called the *father* of v, and the other neighbors are called *children* of v. When terms like child, father, or root are used in a tree, it is always implied that the tree is rooted, that is, it has a root on which these concepts are based. A tree is binary (resp. completely binary) when every internal node has at most (resp. exactly) 2 children. It is easy to prove that every completely binary tree has $n = 2f - 1$ vertices, where f is the number of leaves. A graph G is *bipartite* if $V(G)$ can be partitioned into sets V_1 and V_2 such that every edge of G has one end in V_1 and the other in V_2. Therefore, V_1 and V_2 are independent sets. In Example A.1, the graph is bipartite with $V_1 = \{a, u, w, y\}$ and $V_2 = \{v, x, z\}$.

A graph G is *complete bipartite* if, for every pair of vertices x, y with $x \in V_1$ and $y \in V_2$, it holds that $xy \in E(G)$. We denote by $K_{p,q}$ a complete bipartite graph with p vertices in V_1 and q vertices in V_2. Obviously $K_{p,q}$ has pq edges. It is known that a graph is bipartite if and only if it does not have odd cycles.

A generalization of complete bipartite graph is the definition of *k-partite complete graph*, which is one whose vertex set is partitioned into independent sets $V_1, V_2, \ldots, V_k$ such that there is an edge between two vertices x and y if, and only if, x and y belong to different sets of this partition.

A.5 Hereditary and Monotone Properties

A property of graphs is *monotonic* (resp. *hereditary*) if every subgraph (resp. induced subgraph) of a graph that has the property also has the property. Every monotonic property is hereditary, because if it holds for every subgraph, it also holds for every induced subgraph, but not every hereditary property is monotonic. For example, the property of *having a universal vertex* is neither a hereditary nor a monotonic property. The property of *being triangle-free* is monotonic and, therefore, also hereditary. The property of *being free of induced* C_4 is hereditary, but not monotonic (for example, K_5 is free of induced C_4, but by removing 1 vertex and some edges we can obtain a non-induced C_4 subgraph). In addition, the property of *being a complete graph* is hereditary, but not monotonic.

A.6 Digraphs

A *directed graph* (or *digraph*) D is formed by a triple consisting of a set of vertices $V(D)$, a set of *arcs* $A(D)$ and a function that associates each arc with an *ordered* pair of not necessarily distinct vertices.

For a digraph D, if (u, v) is the pair of vertices associated with the arc $a \in A(D)$, then we say that u and v are the *ends* of a. We also say that a is an arc from u to v, with u being the *tail* of a and v the *head* of a; we also say that u is the *predecessor* of v and v the *successor* of u.

In a digraph D, a *loop* is an arc in D whose ends are the same. We also say that two arcs are *multiple arcs* if they both have the same tail and the same head. A digraph D is said to be *simple* if D does not have multiple arcs.

Note that a simple digraph can have a loop, as well as two arcs, in opposite directions, between the same pair of vertices.

If D is a digraph, then the *underlying graph* of D is the graph G obtained from D by taking the same set of vertices and considering the arcs of D as edges, that is, unordered pairs of vertices.

The notions of walk, trail, path, and cycle in a digraph D are analogous, only respecting the orientation of the arcs. That is, for example, if $P = v_1, a_1, v_2, \ldots, a_{n-1}, v_n$ is a path in D, then the arc a_i is associated with the pair (v_i, v_{i+1}) for every $i \in \{1, \ldots, n-1\}$. It is worth noting that some authors prefer to denote a cycle in a directed graph as a *circuit*. The notions of subgraph and isomorphism are also analogous.

Let v be a vertex in a digraph D. The *out-degree* $d_D^+(v)$ (resp. *in-degree* $d_D^-(v)$) is the number of arcs in which v is the tail (resp. v is the head). The *outgoing neighborhood* (resp. *incoming neighborhood*) of v is the set

$$N_D^+(v) = \{u \mid (v, u) \text{ is associated with an arc of } D\}$$

resp.

$$N_D^-(v) = \{u \mid (u, v) \text{ is associated with an arc of } D\}.$$

When there are no ambiguities, the digraph D can be omitted in the subscript of these definitions.

An *oriented graph* D is the digraph obtained by orienting a simple graph G, that is, by replacing each edge of G with an ordered pair with the same endpoints.

A.7 Planar Graphs

A *curve* is a continuous function $f : [0, 1] \to \mathbb{R}^2$. A u, v-curve is such that $f(0) = u$ and $f(1) = v$ for any $u, v \in \mathbb{R}^2$. A curve is *polygonal* if it is composed of a finite sequence of line segments. A curve f is *simple* if f is injective in the interval $(0, 1)$, that is, the curve itself does not intersect, except possibly in the case where $u = v$. A u, v-curve is *closed* if $u = v$. A *representation or immersion* of a graph G in the plane is a function that associates each vertex with a point in $\mathbb{R}^2$ and each edge $e = uv$ with a u, v–simple polygonal curve. If two edges intersect in the representation of a graph G, such intersection is a *crossing*. A *planar* graph is a graph that admits a representation in the plane $\mathbb{R}^2$ without edge crossings.

A.8 Graph Coloring

A *k-coloring* of a graph $G = (V, E)$ is a function $c : V(G) \to \{1, \ldots, k\}$ such that $c(u) \neq c(v)$ if u and v are neighbors in G. The *chromatic number* of G, denoted by $\chi(G)$, is the smallest natural number k such that G admits a k-coloring. A k-coloring for $k = \chi(G)$ is an *optimal coloring* of G.

If c is a coloring of G, then the set $S_i = \{v \in V(G) \mid c(v) = i\}$ is a *color class* for every $i \in \{1, \ldots, k\}$. Note that each color class defines an independent set of the graph. The following bounds are well known for the chromatic number:

$$\omega(G) \leq \chi(G) \leq \Delta(G) + 1.$$

A.9 Tree and Path Decomposition

A *tree decomposition* (resp. *path decomposition*) of a graph $G = (V, E)$ is a pair $(T, \mathcal{X})$ where T is a tree (resp. a path) and $\mathcal{X}$ is a family of subsets of $V(G)$, so that there is a bijection between $V(T)$ and $\mathcal{X}$, each vertex $t \in V(T)$ being associated with an element, popularly known as *bag*, $B_t \in \mathcal{X}$, satisfying the following conditions:

1. For each $u \in V(G)$, there exists $t \in V(T)$ such that $u \in B_t$.
2. For each edge $e = uv \in E(G)$, there exists $t \in V(T)$ such that $\{u, v\} \subseteq B_t$.
3. For each $u \in V(G)$, if $S = \{t \in V(T) \mid u \in B_t\}$, then $T[S]$ is a subtree of T.

The *width* of the tree (resp. path) decomposition $(T, \mathcal{X})$ of G is the maximum value $\max_{t \in T} |B_t| - 1$. The *treewidth* (resp. *pathwidth*) of G is the smallest width of a tree (resp. path) decomposition of G. The notion of tree width was introduced by Robertson and Seymour (1986) in the proof of the famous graph minor theorem. It is currently a concept widely explored as a structural parameter of the graph in the development of FPT algorithms.

Appendix B
Computational Complexity

B.1 Time Complexity

In a *decision problem*, an instance is given and a YES or NO question is asked about the instance. For example, in the VERTEX COVER decision problem, the instance is a graph G and an integer $k \geq 1$, and the question is whether there exists a subset S with at most k vertices of G such that each edge of G has some endpoint in S.

The *Class* P is defined as the set of decision problems that are *solvable* by a polynomial time algorithm. The *Class* NP is defined as the set of decision problems that are *verifiable* by a polynomial time algorithm. In other words, if we receive a *certificate* (or *proof*) that the instance has a YES answer in the decision problem, it is possible to verify in polynomial time that the *certificate* is valid (with the certificate we have the proof that the instance indeed has a YES answer). As an example, we show below that the VERTEX COVER problem belongs to the class NP.

Theorem B.1 (Garey and Johnson 1979) VERTEX COVER *belongs to* NP.

Proof For a certificate of an instance (G, k), consider a subset S of vertices of the graph G. The verifier algorithm for the certificate S must verify: (a) if $|S| \leq k$ and (b) if every edge of G has at least one endpoint in S. For (a), it is enough to count the elements of S, which takes time $O(k) = O(n)$. For (b), it is enough to go through all the edges of G (there are $O(n^2)$ edges) and check if any of its endpoints is in S ($|S| = k \leq n$), resulting in a total time of $O(kn^2) = O(n^3)$. Therefore, we can verify if S is a cover for G in polynomial time $O(n) + O(n^3) = O(n^3)$. □

A decision problem A *reduces polynomially* to a decision problem B (denoted by $A \leq_p B$) if there exists a polynomial time algorithm F that, for each instance I of A, obtains an instance $F(I)$ of B such that I is YES in A if and only if $F(I)$ is YES in B. The algorithm F is called a *reduction function*. A decision problem B is NP-*hard* if $A \leq_p B$ for every $A \in$ NP, i.e., any problem in NP reduces polynomially to B. A problem is NP-*complete* if it is NP-hard and belongs to NP.

J. Araújo et al., *Introduction to Graph Convexity*, SBMAC Collection on Applied and Computational Mathematics, https://doi.org/10.1007/978-3-031-84128-6

Theorem B.2 (Garey and Johnson 1979) *If a problem B is* NP*-complete and $B \in$* P*, then* P = NP. *Moreover, if* P = NP*, then every* NP*-complete problem is in* P.

The classic question P = NP is one of the most important in computer science, which remains unanswered. It is also one of the seven millennium problems of the Clay Mathematics Institute, with a prize of 1 million dollars for whoever solves it. According to Theorem B.2, one way to solve this question is to show the existence (P = NP) or nonexistence (P $\neq$ NP) of a problem $B \in$ P that is NP-complete.

Another importance of Theorem B.2, besides helping to solve the question P = NP, is that it helps to classify problems by their level of hardness, because, once we show that a problem is NP-complete, we are also showing that this problem does not have a known polynomial time algorithm and that finding this polynomial algorithm is quite difficult, if it exists. We say that a problem in P is *tractable* (or *easy*) to solve while NP-hard problems are *intractable* (or *hard*) to solve.

Next, we show a widely used tool to prove that a problem C is NP-complete.

Theorem B.3 (Garey and Johnson 1979) *If B is* NP*-complete, $C \in$* NP *and $B \leq_p C$, then C is* NP*-complete.*

With Theorem B.3, we can prove that a problem C is NP-complete without directly using the definition. The main objective is then to find an NP-complete problem B and a polynomial reduction from B to C to show that C is NP-complete.

The first proof of NP-hardness was for the SAT problem, defined below. We say that a logical formula is in *conjunctive normal form* if it consists of conjunctions (logical operator *and*, denoted by $\wedge$) of clauses, which, in turn, consist of disjunctions (operator *or*, denoted by $\vee$) of literals (logical variable or complement of logical variable). For example, the logical formula $\phi = (x_1 \vee x_2 \vee x_3) \wedge (\overline{x_1} \vee x_2 \vee \overline{x_3})$, which has 2 clauses with 3 literals each, is in conjunctive normal form, with x_1, x_2, and x_3 being the logical variables. We say that a logical formula in conjunctive normal form is *satisfiable* if there is a true/false assignment to the variables such that the value of the formula with this assignment is true. We have that ϕ is satisfiable, as its value is true when all variables are true.

In the SAT problem, the instance is a formula ϕ in conjunctive normal form and the question is whether ϕ is satisfiable. In the 3SAT problem, the only difference is that the conjunctive normal form formula ϕ has 3 literals per clause. Using Theorem B.3 and the SAT PROBLEM, it is easy to prove that the 3SAT PROBLEM is also NP-complete. Below, we show a proof of NP-completeness from 3SAT.

Theorem B.4 (Garey and Johnson 1979) VERTEX COVER *is* NP*-complete.*

Proof From Theorem B.1, VERTEX COVER is in NP and we know that 3SAT is NP-complete. We obtain a polynomial reduction from 3SAT to VERTEX COVER by constructing an instance (G, k) of VERTEX COVER given an instance ϕ of 3SAT. Let v and c be the number of variables and clauses of ϕ, respectively.

Construction: (a) for each variable x of ϕ, create vertices x and $\overline{x}$ and an edge $x\overline{x}$ in G (we call the vertices x and $\overline{x}$ variable vertices); (b) for each clause of $(x \vee y \vee z)$ of ϕ, create three vertices x, y, z forming a cycle in G (we call these

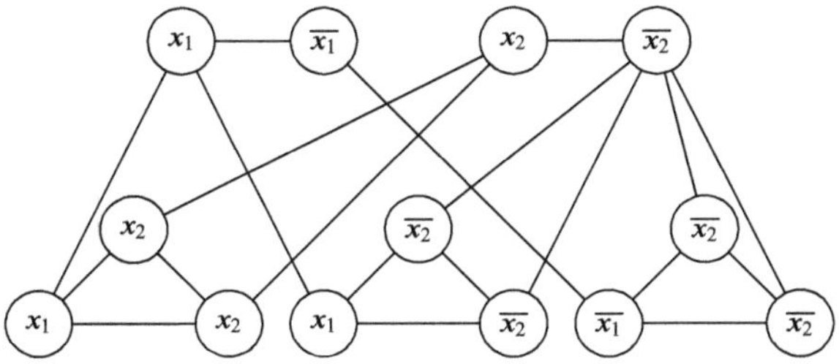

Fig. B.1 $\phi = (x_1 \vee x_2 \vee x_2) \wedge (x_1 \vee \overline{x_2} \vee \overline{x_2}) \wedge (\overline{x_1} \vee \overline{x_2} \vee \overline{x_2})$. The top vertices are related to the variables x_1 and x_2, while each triangle in the bottom represents a clause. The top-down edges connect vertices of corresponding literals

vertices as *clause vertices*); (c) connect the variable vertices to the clause vertices that correspond to the same literal in ϕ; and (d) $k = v + 2c$. Figure B.1 shows the example for $\phi = (x_1 \vee x_2 \vee x_2) \wedge (x_1 \vee \overline{x_2} \vee \overline{x_2}) \wedge (\overline{x_1} \vee \overline{x_2} \vee \overline{x_2})$.

Note that G has $2v + 3c$ vertices and $v + 6c$ edges and is obtained in polynomial time. We show that ϕ is satisfiable if and only if G has a vertex cover of size $k = v + 2c$.

If ϕ is satisfiable, then there is a valid true/false valuation of the variables of ϕ in which each clause is true. Choose the variable vertices of G corresponding to true literals of the valuation and, for each clause C, choose two clause vertices of C so that the third clause vertex corresponds to a true literal. Note that we chose $v + 2c = k$ vertices. Each edge of G has one end that is a chosen vertex: for the edges created in (a), exactly one of the literals x or $\overline{x}$ was chosen; for the edges created in the triangles in (b), two vertices of each triangle were chosen; and for each edge created in (c), one of its ends is a variable vertex and the other is a clause vertex. If the first is not chosen, it corresponds to an false literal and, then, the second is chosen. Thus, the chosen vertices form a vertex cover of size k of G.

Now suppose we have a cover C of size $k \leq v + 2c$ for the graph G (constructed from ϕ). Note that any vertex cover C' of G must cover each edge created in (a) with at least one vertex and each edge of the triangles created (b) with at least two vertices. This implies that each vertex cover of G must have size at least $v + 2c$. Therefore, $v + 2c \leq |C| = k \leq v + 2c$ and, therefore, $|C| = v + 2c$. As C has $v + 2c$ vertices, then C has exactly one vertex for each structure created in (a) and exactly two vertices for each triangle created in (b). For the vertices of C created in (a), give the valuation to their corresponding literal as true and their opposite literal as false. With this, for each clause, there is a clause vertex that is not in the cover C and, therefore, its edge created in (c) must be covered by the variable vertex (for which we valued its literal as true). Then each clause has at least one true literal and this valuation satisfies ϕ. We are done from Theorem B.3. □

B.2 Space Complexity

Analogous to the time complexity, there is the space complexity, related to the amount of memory used by an algorithm, instead of the execution time. Similarly to the classes P, NP, NP-hard, and NP-complete, there are the classes PSPACE, NPSPACE, PSPACE-hard, and PSPACE-complete. However, in space complexity, PSPACE $=$ NPSPACE by Savitch's theorem. It is also known that P $\subseteq$ NP $\subseteq$ PSPACE. The first problem proved PSPACE-complete was QSAT (Quantified SAT) by Stockmeyer and Meyer (1973), defined as follows. A logical formula is given with variables such that every variable is quantified with $\exists$ (exists) or $\forall$ (for all) at the beginning of the formula. The goal is to decide whether the formula is true or false. For example, the formula $\forall x_1 \exists x_2 (x_1 \vee x_2) \wedge (x_1 \vee \overline{x_2})$ is false, because taking x_1 as false, the formula will always be false, regardless of the value of x_2. On the other hand, the formula $\exists x_1 \forall x_2 (x_1 \wedge x_2) \vee (x_1 \wedge \overline{x_2})$ is true, because taking x_1 as true, the formula will always be true, regardless of the value of x_2.

QSAT can be seen as a two-player game, Alice and Bob, who assign values according to the order of quantification of the formula. Alice (resp. Bob) can only assign values to variables quantified with $\exists$ (resp. $\forall$). Alice wins if she makes the formula true, and Bob wins if he makes the formula false.

After the PSPACE-hardness proof of QSAT, several other problems were proved PSPACE-complete, many of them related to two-player games. One of the first is *Node-Kayles*, where two players alternate turns choosing vertices of a graph so that the selected vertices induce an independent set. The last player to be able to play wins the game (obtaining a maximal independent set). In the *Clique Forming game*, the vertices must form a clique, instead of an independent set. It is known that the problems of deciding which player has a winning strategy in Node-Kayles and in Clique Forming game are PSPACE-complete (Schaefer 1978).

B.3 Approximation Complexity

An *optimization problem* is a problem that, given an instance, seeks for a solution to a problem on the instance whose value is optimal (minimum or maximum). That is, every optimization problem P has a type: minimization or maximization. In addition, every instance I of P has a set $\mathrm{Sol}_P(\mathrm{I})$ of solutions and each solution $s \in \mathrm{Sol}_P(\mathrm{I})$ has a value $val_P(s, \mathrm{I})$. The considered instances always have $\mathrm{Sol}_P(\mathrm{I}) \neq \emptyset$, that is, they are feasible instances. *Optimal solution* means a solution with minimum value (resp. maximum) in minimization problems (resp. maximization).

Given an optimization problem P and an instance I, let $opt_P(\mathrm{I})$ be the value of an optimal solution. According to Ausiello et al. (1999), the *approximation ratio* $\mathcal{R}_P(s, \mathrm{I})$ of a solution s of I is defined as

$$\mathcal{R}_P(s, \mathrm{I}) = \max\left\{\frac{opt_P(\mathrm{I})}{val_P(s, \mathrm{I})}, \frac{val_P(s, \mathrm{I})}{opt_P(\mathrm{I})}\right\} \geq 1.$$

Given a function $r(n) \geq 1$, an *r(n)-approximate algorithm* for an optimization problem P is an algorithm that, applied to any instance I of P, produces a solution s of I such that $\mathcal{R}_P(s, \mathrm{I}) \leq r(n)$, where n is the size of the instance I. We say that a problem is $r(n)$-*inapproximable* if it does not have a polynomial $r(n)$-approximate algorithm unless P = NP. We also say that an optimization problem is APX, Log-APX or Poly-APX if it has a polynomial $r(n)$-approximate algorithm for some function $r(n) = k$, $r(n) = O(\log n)$ or $r(n) = O(n^k)$, respectively, where k is a constant. Given optimization problems P_1 and P_2, a reduction from P_1 to P_2 is a pair (f, g) such that, for every fixed rational $r \geq 1$ and every instance I of P_1, (a) $f_r(\mathrm{I})$ is an instance of P_2 computable in polynomial time and, for every solution s of $f_r(\mathrm{I})$, (b) $g_r(\mathrm{I}, S)$ is a solution of I computable in polynomial time.

According to Ausiello et al. (1999), P_1 has an *AP reduction* to P_2 (*approximation preserving* $P_1 \leq_{AP} P_2$) if there exists a reduction (f, g) from P_1 to P_2 and there exists a positive constant γ, such that if $\mathcal{R}_{P_2}(s, f_r(\mathrm{I})) \leq r$, then $\mathcal{R}_{P_1}(g_r(s), \mathrm{I}) \leq 1 + \gamma(r - 1)$ for every rational $r \geq 1$, every instance I of P_1 and every solution s of $f_r(\mathrm{I})$.

We say that an optimization problem P_2 is Poly-APX-hard if $P_1 \leq_{AP} P_2$ for every problem P_1 that is Poly-APX and we say that P_2 is Poly-APX-complete if it is Poly-APX and Poly-APX-hard. APX-hard, APX-complete, Log-APX-hard and Log-APX-complete are defined analogously. It is known, for example, that the *minimum dominating set* problem is Log-APX-complete and that the *maximum clique* problem is Poly-APX-complete (Escoffier and Paschos 2006).

B.4 Parameterized Complexity

With the motivation to deal with NP-hard problems and classify them according to their level of tractability, Downey and Fellows introduced the theory of parameterized complexity. See the book *Parameterized Complexity* by Downey and Fellows (2012) for details. We follow below the notations of the book *Parameterized Complexity Theory* by Flum and Grohe (2006).

A *parameter* k for a computational problem Q is a function that assigns a positive integer $k(x)$ to each instance x of the problem Q. When the instance x of the problem is clear in the context, we can simply write k instead of $k(x)$. A *parameterized problem* is a pair (Q, k), where Q is a decision problem and k is a parameter of the problem Q. Below we show some examples of parameterized problems.

In the parameterized problem 3-COLORING(Δ), the instance is a graph G, the parameter is $\Delta(G)$ (maximum degree of G), and the question is whether there exists a proper vertex coloring of G using at most 3 colors. In the parameterized problem VERTEX COVER(k), the instance is a graph G and an integer k, the parameter is k, and the question is whether there exists $S \subseteq V(G)$, with $|S| \leq k$, in which each edge of G has one end in S. In the parameterized problem CLIQUE(k), the instance is a graph G and an integer k, the parameter is k, and the question is whether there exists $S \subseteq V(G)$, with $|S| \leq k$, in which all vertices of S are adjacent to each other. In the parameterized problem DOMINATING SET(k), the instance is a graph G and an integer k, the parameter is k, and the question is whether there exists $S \subseteq V(G)$, with $|S| \leq k$, in which each vertex of $G - S$ has a neighbor in S.

It is also possible to parameterize in a constant number of parameters, $k_1, k_2, \ldots, k_c$. In this case, it is considered that the parameter k of the parameterized problem is the sum $k = k_1 + \cdots + k_c$. An example is the DOMINANT(Δ, k) problem that has the same instance and the same question as DOMINANT(k), but is parameterized by $\Delta(G)$ and k, that is, it has $\Delta(G) + k$ as a parameter.

Given a computational problem Q, an XP *algorithm* on a parameter k of Q is an algorithm that runs in time $O(f(k) \cdot n^{g(k)})$, where n is the size of the instance of Q, k is the parameter and f and g are computable functions. The *Class* XP is defined as the set of parameterized problems that have XP algorithms. This class contains the parameterized problems that admit a polynomial algorithm in the size of the instance when the parameters are fixed.

Given a computational problem Q, a FPT *algorithm* (fixed parameter tractable) with respect to a parameter k of the problem Q is an algorithm that runs in time $O(f(k) \cdot n^{O(1)})$, where k is the parameter, n is the size of the representation of the instance of the parameterized problem, and f is a computable function. The *Class* FPT is defined as the set of parameterized problems that have FPT algorithms.

We have the analogue of the polynomial reduction of classical complexity theory. Let (Q, k) and (Q', k') be two parameterized problems. An FPT *reduction* from (Q, k) to (Q', k') (denoted by $(Q, k) \leq_{\text{FPT}} (Q', k')$) is an algorithm R that, for every instance x of Q, produces an instance $x' = R(x)$ of Q' such that (a) x is YES in Q if, and only if, x' is YES in Q', (b) there is a computable function g such that $k'(x') \leq g(k(x))$ for every instance x of Q and (c) R is computable by an FPT algorithm (with respect to the parameter k).

Lemma B.1 (Preservation of Fixed Parameter Tractability) *If* $(Q, k) \leq_{\text{FPT}} (Q', k')$ *and* $(Q', k') \in$ FPT, *then* $(Q, k) \in$ FPT.

Two important classes of parameterized problems are the *Classes* W[1] and W[2]. The original definition of these classes depends on the definition of boolean circuits and, for this, several additional notations are necessary, which we will leave for the next subsection. We initially provide simpler equivalent definitions (as in (Flum and Grohe 2006)). Let W[1] be the class of parameterized problems that have an FPT reduction to the CLIQUE(k) problem. Let W[2] be the class of parameterized problems that have an FPT reduction to the DOMINANT(k) problem.

Just as P $\subseteq$ NP in classical complexity, we have that FPT $\subseteq$ W[1] $\subseteq$ W[2] $\subseteq$ XP in parameterized complexity.

For $t \in \{1, 2\}$, we say that a parameterized problem (Q', k') is W[t]-hard if for every problem $(Q, k) \in$ W[t], we have that $(Q, k) \leq_{\text{FPT}} (Q', k')$. In addition, (Q', k') is W[t]-complete if it is in W[t] and is W[t]-hard. Just as there is the conjecture P $\neq$ NP in classical complexity theory, there is also the conjecture FPT $\neq$ W[1] $\neq$ W[2] in parameterized complexity. It is known that if a W[1]-complete problem is in FPT, then FPT = W[1] and, if a W[2]-complete problem is in W[1], then W[1] = W[2]. It is also known that VERTEX COVER(k) belongs to FPT and that, by the definition given above of the classes W[1] and W[2], the problem CLIQUE(k) is W[1]-complete and the problem DOMINATING SET(k) is W[2]-complete.

W Hierarchy

To provide the original definition of the classes W[1] and W[2], mentioned in the previous section, it is necessary to first define a *boolean circuit*, which is a directed acyclic graph in which the vertices are labeled as follows: (a) every vertex with in-degree 0 is an *input node*, (b) every vertex with in-degree 1 is a *negation node*, and (c) every vertex with in-degree at least 2 is a *conjunction node* or *disjunction node*. The terms negation, conjunction, and disjunction are associated with the classical logical operators $\neg$, $\wedge$, and $\vee$, respectively. In addition, there is only one vertex with out-degree 0 that is also labeled as an *output node*. We say that a node is *large* if it has in-degree at least 3, i.e., it can only be a conjunction or disjunction node.

For simplicity, as the boolean circuit is acyclic, it is usually represented without the orientations of the edges, assuming the natural direction from the input nodes to the output node, from top to bottom, as exemplified in Fig. B.2.

The *depth* of a boolean circuit is the maximum size of a path from an input node to the output node. The *weft* of a boolean circuit is the maximum number of large nodes in a path from an input node to the output node.

Assigning 0–1 (true–false) values to the input nodes determines the value of each node in the circuit in the expected way, applying the logical operators. With the term

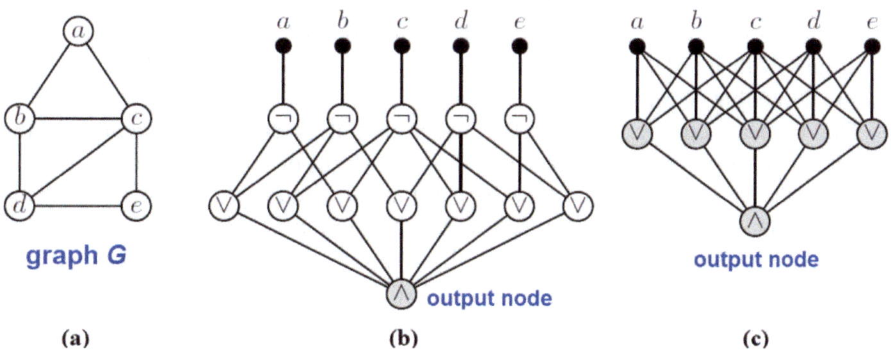

Fig. B.2 Examples of boolean circuits to decide whether a set of vertices is independent or dominant in a graph G. Large nodes (which enter the count of the interlacing) are in gray: (**a**) Graph G with 5 vertices and 7 edges; (**b**) boolean circuit of depth 3 and interlacing 1, only satisfied by *independent sets* of G (each disjunction node corresponds to an edge of G and has input degree 2); (**c**) circuit of depth 2 and interlacing 2, only satisfied by *dominant sets* of G (each disjunction node corresponds to the closed neighborhood of a vertex of G)

assignment, we will be referring to an assignment of 0–1 values to the input nodes of a given boolean circuit. We say that an assignment satisfies the boolean circuit if the value of the output node is 1.

Figure B.2b, c shows boolean circuits associated with the independent set and dominating set problems of the graph G in Fig. B.2a, respectively. In Fig. B.2b, each black vertex is connected to only one "$\neg$" vertex and both are referring to a vertex of G, while each "$\vee$" vertex is referring to an edge of G, being connected to the two "$\neg$" vertices referring to its endpoints; and the "$\wedge$" vertex is connected to the "$\vee$" vertices. Note in Fig. B.2b that if we take vertices a and b as 1 (true), the first "$\vee$" vertex (from left to right) will be 0 (false) and, therefore, the output vertex "$\wedge$" will be 0. As this occurs for each pair of adjacent vertices of G, then, for the output vertex to be 1, all input vertices with 1 must correspond to an independent set in G. To obtain a boolean circuit for the problem CLIQUE(k), just take the circuit of the independent set related to the complement of the graph G in Fig. B.2a. In Fig. B.2c, each black vertex corresponds to a vertex of G, the "$\vee$" vertices correspond to the closed neighborhood of a vertex of G, and the "$\wedge$" vertex is connected to the "$\vee$" vertices. To have 0 at the output vertex in the circuit of Fig. B.2c, then at least one "$\vee$" vertex must have all its input vertices 0, that is, there must exist a closed neighborhood of some vertex of G with all its vertices 0. Thus, to have 1 at the output vertex, there must be at least one input vertex 1 in the closed neighborhood of each vertex of G, that is, these input vertices 1 form a dominating set of G.

Deciding whether a boolean circuit is satisfiable, that is, if there is an assignment that satisfies it, is an NP-complete problem, because the 3SAT Problem is a particular case: every instance of 3SAT can be represented in a boolean circuit of depth ≤ 3 and interlacing ≤ 2 (conjunction of disjunctions).

Let *WCD* (*weighted circuit satisfiability*) be the problem of deciding whether, given a boolean circuit C and an integer k, there is an assignment that satisfies C with exactly k values equal to 1. Let W[t], for $t \geq 1$ integer, be the class of parameterized problems that admit an FPT reduction to the WCD problem restricted to circuits with interlacing t, parameterized by k (number of 1 values in the input nodes). Note that $\mathrm{W}[t] \subseteq \mathrm{W}[t+1]$ for all $t \geq 1$.

The boolean circuits of Fig. B.2 show that it is possible to represent the independent set problem (consequently the CLIQUE(k) problem) in boolean circuits with interlacing 1 and the DOMINANT(k) problem in circuits with interlacing 2. That is, CLIQUE(k) and DOMINANT(k) are in W[1] and in W[2] respectively.

It is known that CLIQUE(k) is W[1]-hard (every problem in W[1] has an FPT reduction to it) and that DOMINANT(k) is W[2]-hard (every problem in W[2] has an FPT reduction to it).

References

Albenque, M., and K. Knauer. 2016. Convexity in partial cubes: The hull number. *Discrete Mathematics* 339(2):866–876.
Alcón, L., B. Brešar, T. Gologranc, M. Gutierrez, T. Kraner Šumenjak, I. Peterin, and A. Tepeh. 2015. Toll convexity. *European Journal of Combinatorics* 46:161–175.
Amusements in Mathematics. 1917. *Nature* 100:302–303.
Anand, B.S., U. Chandran S.V., M. Changat, S. Klavžar, and E.J. Thomas. 2019. Characterization of general position sets and its applications to cographs and bipartite graphs. *Applied Mathematics and Computation* 359:84–89.
Anand, B.S., U. Chandran S.V., M. Changat, M. C. Dourado, F.H. Nezhad, and P.N. Shenoi. 2020. On the Carathéodory and exchange numbers of geodetic convexity in graphs. *Theoretical Computer Science* 804:46–57.
Araújo, J., and P.S.M. Arraes. 2022. Hull and geodetic numbers for some classes of oriented graphs. *Theoretical Computer Science* 323:14–27.
Araújo, R., and R. Sampaio. 2023. Domination and convexity problems in the target set selection model. *Theoretical Computer Science* 330:14–23.
Araújo, R., R. Sampaio, and J.L. Szwarcfiter. 2013. The convexity of induced paths of order three. *Electronic Notes in Discrete Mathematics* 44:109–114.
Araújo, R., R. Sampaio,V. F. dos Santos, and J.L. Szwarcfiter. 2018. The convexity of induced paths of order three and applications: Complexity aspects. *Theoretical Computer Science* 237:33–42.
Araújo, J., A.K. Maia, P.P. Medeiros, and L. Penso. 2023. On the hull and interval numbers of oriented graphs.
Araújo, S. N., J.M. Brito, R. Folz, R. de Freitas, and R.M. Sampaio. 2024a. Graph convexity impartial games: Complexity and winning strategies. *Theoretical Computer Science* 998:114534.
Araújo, S.N., J.M. Brito, R. Folz, R. de Freitas, and R.M. Sampaio. 2024b. Graph convexity partizan games. In: 30th *International Computing and Combinatorics Conference (COCOON'24).*
Araújo, J., V. Campos, F. Giroire, N. Nisse, L. Sampaio, and R. Soares. 2013. On the hull number of some graph classes. *Theoretical Computer Science* 475:1–12.
Araújo, J., M.C. Dourado, F. Protti, and R. Sampaio. 2025. The iteration time and the general position number in graph convexities. *Applied Mathematics and Computation* 487:129084.
Ausiello, G., M. Protasi, A. Marchetti-Spaccamela, G. Gambosi, P. Crescenzi, and V. Kann. 1999. *Complexity and approximation: Combinatorial optim problems and their approximability properties*, 524. Berlin: Springer.
Bárány, I. 2021. *Combinatorial convexity*, Vol. 77. University Lecture Series, viii+148. Providence: American Mathematical Society.
Bandelt, H.-J. 1989. Graphs with intrinsic S_3 convexities. *Journal of Graph Theory* 13(2):215–228.

J. Araújo et al., *Introduction to Graph Convexity*, SBMAC Collection on Applied and Computational Mathematics, https://doi.org/10.1007/978-3-031-84128-6

Bandelt, H.-J., and V. Chepoi. 1996. A Helly theorem in weakly modular space. *Discrete Mathematics* 160(1):25–39.

Bandelt, H.-J., and H.M. Mulder. 1990. Helly theorems for dismantlable graphs and pseudo-modular graphs. In: *Topics in combinatorics and graph theory: Essays in honour of Gerhard Ringel*. ed. R. Bodendiek and R. Henn, 65–71. Heidelberg: Physica-Verlag HD.

Barbosa, R.M., E.M.M. Coelho, M.C. Dourado, D. Rautenbach, and J.L. Szwarcfiter. 2012. On the Carathéodory number for the convexity of paths of order three. *SIAM Journal on Discrete Mathematics* 26(3):929–939.

Bazgan, C., M. Chopin, A. Nichterlein, and F. Sikora. 2014. Parameterized inapproximability of target set selection and generalizations. *Computability* 3:135–145.

Ben-Zwi, O., D. Hermelin, D. Lokshtanov, and I. Newman. 2011. Treewidth governs the complexity of target set selection. *Discrete Optimization* 8:87–96.

Benesh, B.J., D.C. Ernst, M. Meyer, S. Salmon, and N. Sieben. 2024. Impartial geodetic building games on graphs. arXiv:2307.07095.

Benevides, F., and M. Przykucki. 2013. On slowly percolating sets of minimal size in bootstrap percolation. *The Electronic Journal of Combinatorics* 20(2):P46.

Benevides, F., V. Campos, M. C. Dourado, R. Sampaio, and A. Silva. 2015. The maximum time of 2-neighbour bootstrap percolation: Algorithmic aspects. *European Journal of Combinatorics* 48:88–99.

Benevides, F., V. Campos, M. C. Dourado, R. Sampaio, and A. Silva. 2016. The maximum infection time in the geodesic and monophonic convexities. *Theoretical Computer Science* 609:287–295.

Berge, C., and P. Duchet. 1975. Une généralisation du théorème de Gilmore. *Cahiers du Centre d'études de recherche operationnelle* 17:117–123.

Bondy, J.A., and U.S.R. Murty. 2008. *Graph theory*, xii+663. London: Springer.

Brandstädt, A., V.B. Le, and J.P. Spinrad. 1999. *Graph classes: A survey*, xi+295. SIAM.

Brosse, C., N. Martins, N. Nisse, and R. Sampaio. 2025. The convex set forming game. INRIA, hal-04710504.

Bryant, V.W., J.E. Dawson, and H. Perfect. 1978. Hereditary circuit spaces. *Compositio Mathematica* 37(3):339–351.

Buckley, F., and F. Harary. 1985a. Closed geodetic games for graphs. *Congressus Numerantium* 47. In *Proceedings of 16th Southeastern Conference on Combinatorics, Graph Theory and Computing*, 131–138.

Buckley, F., and F. Harary. 1985b. Geodetic games for graphs. *Quaestiones Mathematicae* 8:321–334.

Bueno, L.R., L. Penso, F. Protti, V.R. Ramos, D. Rautenbach, and U.S. Souza. 2018. On the hardness of finding the geodetic number of a subcubic graph. *Information Processing Letters* 135:22–27.

Cáceres, J., A. Márquez, and M.L. Puertas. 2008. Steiner distance and convexity in graphs. *European Journal of Combinatorics* 29(3):726–736.

Calder, J.R. 1971. Some Elementary Properties of Interval Convexities. *Journal of the London Mathematical Society* s2–3(3):422–428.

Campos, V., R. Sampaio, A. Silva, and J.L. Szwarcfiter. 2015. Graphs with few P4's under the convexity of paths of order three. *Theoretical Computer Science* 192:28–39.

Canoy Jr., S.R., and I.J.L. Garces. 2002. Convex sets under some graph operations. *Graphs and Combinatorics* 18(4):787–793.

Cappelle, M.R., E.M.M. Coelho, H. Coelho, B.R. Silva, U.S. Souza, and F. Protti. 2022. P3-convexity on graphs with diameter two: Computing hull and interval numbers. *Theoretical Computer Science* 321:368–378.

Carathéodory, C. 1911. Über den variabilitätsbereich der fourier'schen konstanten von positiven harmonischen funktionen. *Rendiconti del Circolo Matematico di Palermo, Series 1* 32(1):193–217.

Carvalho, M.T. 2016. O Número de Helly na Convexidade Geodética em Grafos. PhD thesis. UFRJ.

Castonguay, D., E.M.M. Coelho, H. Coelho, and J.R. Nascimento. 2019. A note on the convexity number for complementary prisms. *Discrete Mathematics & Theoretical Computer Science* 21.

Centeno, C.C., M.C. Dourado, and J.L. Szwarcfiter. 2009. On the convexity of paths of length two in undirected graphs. *Electronic Notes in Discrete Mathematics* 32:11–18. DIMAP Workshop on Algorithmic Graph Theory.

Centeno, C.C., M.C. Dourado, L. Penso, D. Rautenbach, and J.L. Szwarcfiter. 2011. Irreversible conversion of graphs. *Theoretical Computer Science* 412(29):3693–3700.

Cepoĭ, V.D. 1986. Some properties of d-convexity in triangulated graphs (in Russian). *Matematiceskie issledovanija* 87:164–177.

Chakraborty, D., S. Das, F. Foucaud, H. Gahlawat, D. Lajou, and B. Roy. 2020a. Algorithms and complexity for geodetic sets on planar and chordal graphs. In *ISAAC 2020*. LIPIcs, 7:1–7:15.

Chakraborty, D., F. Foucaud, H. Gahlawat, S.K. Ghosh, and B. Roy. 2020b. Hardness and approximation for the geodetic set problem in some graph classes. *Algorithms and Discret. Appl. Math.. 6th international conference, CALDAM 2020, Hyderabad, India, February 13–15, 2020*. ed. M. Changat and S. Das. Vol. 12016, 102–115. Lecture Notes in Computer Science. Berlin: Springer .

Chalupa, J., P.L. Leath, and G.R. Reich. 1979. Bootstrap percolation on a Bethe lattice. *Journal of Physics C: Solid State Physics* 12(1):31–35.

Chang, G.J., L.-D. Tong, and H.-T. Wang. 2004. Geodetic spectra of graphs. *European Journal of Combinatorics* 25(3):383–391.

Changat, M., and J. Mathews. 1999. On triangle path convexity in graphs. *Discrete Mathematics* 206:91–95.

Changat, M., S. Klavzar, and H.M. Mulder. 2001. The all-paths transit function of a graph. *Czechoslovak Mathematical Journal* 51(2):439–448.

Changat, M., H.M. Mulder, and G. Sierksma. 2005. Convexities related to path properties on graphs. *Discrete Mathematics* 290(2–3):117–131.

Changat, M., G.N. Prasanth, and J. Mathews. 2009. Triangle path transit functions, betweenness and pseudo-modular graphs. *Discrete Mathematics* 309(6):1575–1583.

Chartrand, G., and P. Zhang. 1999. Convex sets in graphs. *Congressus Numerantium* 19–32.

Chartrand, G., and P. Zhang. 2000. The geodetic number of an oriented graph. *European Journal of Combinatorics* 21(2):181–189.

Chartrand, G., F. Harary, and P. Zhang. 2000. On the hull number of a graph. *Ars Combinatoria* 57:129–138.

Chartrand, G., J.F. Fink, and P. Zhang. 2002a. Convexity in oriented graphs. *Discrete Applied Mathematics* 116(1–2):115–126.

Chartrand, G., C.E. Wall, and P. Zhang. 2002b. The convexity number of a graph. *Graphs and Combinatorics* 18(2):209–217.

Chartrand, G., C.E. Wall, and P. Zhang. 2002c. On the geodetic number of a graph. *Networks* 39(1):1–6.

Chartrand, G., J.F. Fink, and P. Zhang. 2003. The hull number of an oriented graph. *International Journal of Mathematics* 2003(36):2265–2275.

Chen, N. 2009. On the approximability of influence in social networks. *SIAM Journal on Discrete Mathematics* 23(3):1400–1415.

Chopin, M., A. Nichterlein, R. Niedermeier, and M. Weller, 2014. Constant thresholds can make target set selection tractable. *Theory of Computing Systems* 55(1):61–83.

Cicalese, F., M. Milanič, and U. Vaccaro. 2013. On the approximability and exact algorithms for vector domination and related problems in graphs. *Theoretical Computer Science* 161(6):750–767.

Cicalese, F., G. Cordasco, L. Gargano, M. Milanič, and U.Vaccaro. 2014. Latency bounded target set selection in social networks. *Theoretical Computer Science* 535:1–15.

Coelho, E.M.M., M.C. Dourado, D. Rautenbach, and J.L. Szwarcfiter. 2014. The Carathéodory number of the P3 convexity of chordal graphs. *Theoretical Computer Science* 172:104–108.

Coelho, E.M.M., M. C. Dourado, and R. Sampaio. 2015. Inapproximability results for graph convexity parameters. *Theoretical Computer Science* 600:49–58 (cit. on pp. 20, 22, 53, 71).

Coelho, E.M.M., H. Coelho, J.R. Nascimento, and J.L. Szwarcfiter. 2022. A polynomial time algorithm for geodetic hull number for complementary prisms. *RAIRO-Theoretical Informatics and Applications* 56(1):1–11.

Corneil, D.G., H. Lerchs, and L.S. Burlingham. 1981. Complement reducible graphs. *Theoretical Computer Science* 3(3):163–174.

Costa, E.R., M.C. Dourado, and R. Sampaio. 2015. Inapproximability results related to monophonic convexity. *Theoretical Computer Science* 197:70–74. Distance Geometry and Applications.

Costa, E.R., V.L. Pessoa, R. Sampaio, and R. Soares. 2020. PSPACEcompleteness of two graph coloring games. *Theoretical Computer Science* 824–825:36–45.

da Silva, A.R. 2014. O número de Helly na convexidade geodética: contribuições teóricas e práticas". MA Thesis. UFRJ.

Dailly, A., H. Gahlawat, and Z.M. Myint. 2024. The Closed Geodetic Game: Algorithms and strategies. arXiv:2409.20505.

Dong, L., C. Lu, and X. Wang. 2009. The upper and lower geodetic numbers of graphs. *Ars Combinatoria* 91.

Dourado, M.C., and A. R. da Silva. 2017. Inapproximability results and bounds for the Helly and Radon numbers of a graph. *Theoretical Computer Science* 232:134–141.

Dourado, M.C. and R. Sampaio. 2016. Complexity aspects of the triangle path convexity. *Theoretical Computer Science* 206:39–47.

Dourado, M.C., J.G. Gimbel, J. Kratochvíl, F. Protti, and J.L. Szwarcfiter. 2009. On the computation of the hull number of a graph. *Discrete Mathematics* 309(18):5668–5674. Combinatorics 2006, A Meeting in Celebration of Pavol Hell's 60th Birthday (May 1–5, 2006) .

Dourado, M.C., F. Protti, and J.L. Szwarcfiter. 2010. Complexity results related to monophonic convexity. *Theoretical Computer Science* 158(12):1268–1274.

Dourado, M.C., F. Protti, D. Rautenbach, and J.L. Szwarcfiter. 2010a. On the hull number of triangle-free graphs. *SIAM Journal on Discrete Mathematics* 23(4):2163–2172.

Dourado, M.C., F. Protti, D. Rautenbach, and J.L. Szwarcfiter. 2010b. Some remarks on the geodetic number of a graph. *Discrete Mathematics* 310(4):832–837.

Dourado, M.C., F. Protti, D. Rautenbach, and J.L. Szwarcfiter. 2012. On the convexity number of graphs. *Graphs and Combinatorics* 28(3):333–345.

Dourado, M.C., D. Rautenbach,V.F. dos Santos, P.M. Schäfer, J. L. Szwarcfiter, and A. Toman. 2013a. Algorithmic and structural aspects of the P3-Radon number. *Annals of Operations Research* 206:75–91.

Dourado, M.C., D. Rautenbach, V.G.P. de Sá, and J.L. Szwarcfiter. 2013b. On the geodetic Radon number of grids. *Discrete Mathematics* 313(1):111–121.

Dourado, M.C., D. Rautenbach, V.F. dos Santos, P.M. Schäfer, and J.L. Szwarcfiter. 2013c. On the Carathéodory number of interval and graph convexities. *Theoretical Computer Science* 510:127–135.

Dourado, M.C., R.A. Oliveira, and F. Protti. 2014. Algorithmic aspects of Steiner convexity and enumeration of Steiner trees. *Annals of Operations Research* 223(1):155–171.

Dourado, M.C., V.G.P. de Sá, D. Rautenbach, and J.L. Szwarcfiter. 2016a. Near linear-time algorithm for the geodetic Radon number of grids. *Discrete Applied Mathematics* 210:277–283.

Dourado, M.C., R.A. Oliveira, F. Protti, and D. Rautenbach. 2016b. On the geodetic iteration number of distance-hereditary graphs. *Discrete Mathematics* 339(2):489–498.

Dourado, M.C., L. Penso, and D. Rautenbach. 2016c. On the geodetic hull number of PK-free graphs. *Theoretical Computer Science* 640:52–60.

Dourado, M.C., L. Penso, and D. Rautenbach. 2017. Geodetic convexity parameters for $(q, q-4)$-graphs. *Theoretical Computer Science* 223:64–71.

Dourado, M.C., V.S. Ponciano, and R.L.O. da Silva. 2022a. The hull number in the convexity of induced paths of order 3. *Theoretical Computer Science* 906:52–63.

Dourado, M.C., V.S. Ponciano, and R.L.O. da Silva. 2022a. On the monophonic rank of a graph. *Discrete Mathematics & Theoretical Computer Science* 24(2).

Dourado, M.C., M. Gutierrez, F. Protti, R. Sampaio, and S. Tondato. 2025. Characterizations of graph classes via convex geometries: A survey. *Discrete Applied Mathematics* 360:246–257.

Downey, R., and M. Fellows. 2012. *Parameterized complexity*, xv+533. Berlin: Springer (cit. on pp. 55, 96, 121).

Dragan, F.F., F. Nicolai, and A. Brandstädt. 1999. Convexity and HHD-free graphs. *SIAM Journal on Discrete Mathematics* 12(1):119–135.

Dreyer, P.A., and F.S. Roberts. 2009. Irreversible k-threshold processes: Graphtheoretical threshold models of the spread of disease and of opinion. *Theoretical Computer Science* 157:1615–1627.

Duchet, P. 1987. Convexity in combinatorial structures. *Proceedings of the 14th Winter School on Abstract Analysis*. Proceedings of the 14th Winter School on Abstract Analysis. Circolo Matematico di Palermo, 261–293.

Duchet, P. 1988. Convex sets in graphs, II. Minimal path convexity. *Journal of Combinatorial Theory, Series B* 44(3):307–316.

Dudeney, H.E. 1908. *The Canterbury Puzzles and other curious problems*, 258. New York: E. P. Dutton.

Dudeney, H.E. 1917. *Amusements in mathematics*, v+259. London, Edinburgh and New York: Thomas Nelson and Sons.

Edelman, P.H. and R.E. Jamison. 1985. The theory of convex geometries. *Geometriae Dedicata* 19:247–270.

Ehard, S., and D. Rautenbach. 2019. On some tractable and hard instances for partial incentives and target set selection. *Discrete Optimization* 34:100547.

Eirinaki, M., N. Moniz, and K. Potika. 2016. Threshold-bounded influence dominating sets for recommendations in social networks. In *IEEE BD-Cloud-SocialCom-SustainCom*, 408–415.

Ekim, T., A. Erey, P. Heggernes, P. van't Hof, and D. Meister. 2012. Computing minimum geodetic sets of proper interval graphs. In *LATIN 2012*, 279–290. Berlin: Springer.

Erdős, P., and G. Szekeres. 1935. A combinatorial problem in geometry. *Compositio Mathematica* 2:463–470.

Erdős, P., E. Fried, A. Hajnal, and E.C. Milner. 1972. Some remarks on simple tournaments. *Algebra universalis* 2:238–245.

Ertem, Z., E. Lykhovyd, Y. Wang, and S. Butenko. 2020. The maximum independent union of cliques problem: Complexity and exact approaches. *Journal of Global Optimization* 76:545–562.

Escoffier, B., and V.T. Paschos. 2006. Completeness in approximation classes beyond APX. *Theoretical Computer Science* 359(1):369–377.

Everett, M.G. and S.B. Seidman. 1985. The hull number of a graph. *Discrete Mathematics* 57(3):217–223.

Farber, M. 1983. Characterizations of strongly chordal graphs. *Discrete Mathematics* 43:173–189.

Farber, M., and R.E. Jamison. 1986. Convexity in graphs and hypergraphs. *SIAM Journal on Discrete Mathematics* 7(3):433–444.

Farrugia, A. 2005. Orientable convexity, geodetic and hull numbers in graphs. *Theoretical Computer Science* 148(3):256–262.

Flammenkamp, A. 1998. Progress in the no-three-in-line problem, II. *Journal of Combinatorial Theory, Series A* 81(1):108–113.

Flocchini, P., R. Královič, P. Ružička, A. Roncato, and N. Santoro. 2003. On time versus size for monotone dynamic monopolies in regular topologies. *Journal of Discrete Algorithms* 1(2):129–150.

Flum, J., and M. Grohe. 2006. *Parameterized complexity theory*, xiii+495. New York: Springer.

Garey, M.R. and D.S. Johnson. 1979. *Computers and intractability: A guide to the theory of NP-completeness*, x+340. W. H. Freeman: New York (cit. on pp. 66, 72, 83, 117, 118).

Ghorbani, M., S. Klavžar, H.R. Maimani, M. Momeni, F. Rahimi-Mahid, and G. Rus. 2021. The general position problem on Kneser graphs and on some graph operations. *Discussiones Mathematicae Graph Theory* 41:1199–1213.

Gimbel, J.G. 2003. Some remarks on the convexity number of a graph. *Graphs and Combinatorics* 19(3):357–361.

Golumbic, M.C. 1980. *Algorithmic graph theory and perfect graphs*, xx+284. New York: Academic Press.

Gruber, P.M., and J.M. Wills. 1993. *Handbook of convex geometry*, xi+735. Amsterdam: North Holland .

Gutierrez, M., F. Protti, and S. Tondato. 2023. Convex geometries over induced paths with bounded length. *Disc Mathematics* 346(1):113133.

Gutin, G., and A. Yeo. 2009. On the number of connected convex subgraphs of a connected acyclic digraph. *Theoretical Computer Science* 157(7):1660–1662.

Haglin, D.J. and M.J. Wolf. 1996. On Convex Subsets in Tournaments. *SIAM Journal on Discrete Mathematics* 9(1):63–70.

Harary, F. 1984. Convexity in graphs: Achievement and avoidance games. In *Convexity and graph theory*. ed. M. Rosenfeld and J. Zaks. Vol. 87, 323. North-Holland Mathematics Studies. Amsterdam: North-Holland.

Harary, F., and J. Nieminem. 1981. Convexity in graphs. *Journal of Differential Geometry* 16(1):185–190.

Harary, F., E. Loukakis, and C. Tsouros. 1993. The geodetic number of a graph. *Mathematical and Computer Modelling* 17(11):89–95.

Haynes, T.W., M.A. Henning, and C. Tiller. 2003. Geodetic achievement and avoidance games for graphs. *Quaestiones Mathematicae* 26:389–397.

Haynes, T.W., S.T. Hedetniemi, and M.A. Henning. 2020. *Topics in domination in graphs*, viii+545. Developments in Mathematics. Berlin: Springer.

Helly, E. 1923. Über Mengen konvexer Körper mit gemeinschaftlichen Punkten. *Jahresbericht der Deutschen Mathematiker-Vereinigung* 32:175–176.

Holroyd, A. 2003. Sharp metastability threshold for two-dimensional bootstrap percolation. *Probability Theory and Related Fields* 125:195–224.

Howorka, E. 1981. A characterization of ptolemaic graphs. *Journal of Graph Theory* 5:323–331.

Hung, J.-T., L.-D. Tong, and H.-T. Wang. 2009. The hull and geodetic numbers of orientations of graphs. *Discrete Mathematics* 309(8):2134–2139.

Jamison, R.E. 1974. A general theory of convexity. PhD Thesis. University of Washington.

Jamison, R.E. 1981. Partition numbers for trees and ordered sets. *Pacific Journal of Mathematics* 96(1):115–140.

Jamison, R.E. 1982. A perspective on abstract convexity: Classifying alignments by varieties. *Convexity and Related Combinatorial Geometry*.

Jamison, R. E., and R. Nowakowski. 1984. A Helly theorem for convexity in graphs. *Discrete Mathematics* 51(1):35–39.

Jamison, B., and S. Olariu. 1988. On the semi-perfect elimination. *Advances in Applied Mathematics* 9:364–376.

Kanté, M.M., and L. Nourine. 2013. Polynomial time algorithms for computing a minimum hull set in distance-hereditary and chordal graphs. In *SOFSEM 2013*, 268–279. Berlin: Springer.

Kanté, M.M., R. Sampaio, V.F. dos Santos, and J.L. Szwarcfiter. 2017. On the geodetic rank of a graph. *Journal of Combinatorics* 8(2):323–340.

Kanté, M., T. Marcilon, and R. Sampaio. 2019. On the parameterized complexity of the geodesic hull number. *Theoretical Computer Science* 791:10–27.

Karp, R. 1972. Reducibility among combinatorial problems. *Complexity of computer computations*, 85–103. New York: Plenum.

Keiler, L., C.V.G.C. Lima, A.K. Maia, R. Sampaio, and I. Sau. 2023. Target set selection with maximum activation time. *Theoretical Computer Science* 338:199–217.

Kellerhals, L., and T. Koana. 2022. Parameterized complexity of geodetic set. *Journal of Graph Algorithms and Applications* 26(4):401–419.

Kempe, D., J.M. Kleinberg, and É. Tardos. 2003. Maximizing the spread of influence through a social network. In *Proceedings of the 9th ACM SIGKDD*. KDD '03, 137–146.

Kim, B.K. 2004. A lower bound for the convexity number of some graphs. *Journal of Applied Mathematics and Computing* 14:185–191.

Klavžar, S., Neethu P.K., and U. Chandran S.V. 2022. The general position achievement game played on graphs. *Theoretical Computer Science* 317:109–116 (cit. on p. 103).

Krein, M., and D. Milman. 1940. On extreme points of regular convex sets. *Studia Mathematica* 9:133–138.

Leimer, H.-G. 1993. Optimal decomposition by clique separators. *Discrete Mathematics* 113:99–123.

Lekkerkerker, C., J. Boland. 1962. Representation of a finite graph by a set of intervals on the real line. *Fundamenta Mathematicae* 51(1):45–64.

Levi, F.W. 1951. On Helly's theorem and the axioms of convexity. *The Journal of Indian Mathematical Society* 15:65–76.

Lira, E.S. 2016. O número de Carathéodory na convexidade geodésica de grafos. MA Thesis. Universidade Federal de Goiás.

Lund, C., and M. Yannakakis. 1994. On the hardness of approximating minimization problems. *Journal of the ACM* 41(5):960–981.

Mafort, R., and F. Protti. 2020. Vector Domination in split-indifferent graphs. *Information Processing Letters* 155:105899.

Manuel, P., and S. Klavžar. 2018. A general position problem in graph theory. *Bulletin of the Australian Mathematical Society* 98(2):177–187.

Marcilon, T., and R. Sampaio. 2018a. The maximum infection time of the P3 convexity in graphs with bounded maximum degree. *Theoretical Computer Science* 251:245–257.

Marcilon, T., and R. Sampaio. 2018b. The maximum time of 2-neighbor bootstrap percolation: Complexity results. *Theoretical Computer Science* 708:1–17.

Marcilon, T., and R. Sampaio. 2018c. The P3 infection time is W[1]-hard parameterized by the treewidth. *Information Processing Letters* 132:55–61.

McConnell, R.M., and J.P. Spinrad. 1999. Modular decomposition and transitive orientation. *Discrete Mathematics* 201(1):189–241.

Mezzini, M. 2018. Polynomial time algorithm for computing a minimum geodetic set in outerplanar graphs. *Theoretical Computer Science* 745:63–74.

Minkowski, H. 1903. Volumen und Oberfläche. *Mathematische Annalen* 57:447–495.

Minkowski, H. 1911. Theorie der Konvexen Korper. *Gesammelte Abhandlungen*, ed. D. Hilbert. Vol. 2, 131–229. Capítulo 10. Leipzig: Teubner.

Moon, J.W. 1972. Embedding tournaments in simple tournaments. *Discrete Mathematics* 2(4):389–395.

Moran, S., and A. Yehudayoff. 2020. On weak ε-nets and the Radon number. *Discrete & Computational Geometry* 64(4):1125–1140.

Moscarini, M. 2020. On the geodetic iteration number of a graph in which geodesic and monophonic convexities are equivalent. *Theoretical Computer Science* 283:142–152.

Nečásková, M. 1988. A note on the achievement geodetic games. *Quaestiones Mathematicae* 12:115–119.

Neethu P.K., and U. Chandran S.V. 2022. A note on the convexity number of the complementary prisms of trees. *Theoretical Computer Science* 319:480–486 (cit. on p. 76).

Nichterlein, A., R. Niedermeier, J. Uhlmann, and M. Weller. 2013. On tractable cases of target set selection. *SocialNetwork Analysis and Mining* 3(2):233–256.

Nordhaus, E.A. and J.W. Gaddum. 1956. On complementary graphs. *The American Mathematical Monthly* 63(3):175–177.

Olariu, S. 1989. Weak bipolarizable graphs. *Discrete Mathematics* 74:159–171 (cit. on p. 38).

Padmavathi, S.V. 2015. Relation between convexity number and independence number of a graph. *Annals of Pure and Applied Logic* 9(1):9–12.

Parker, D.B., and R.F. Westhoff. 2012. Convex invariants in multipartite tournaments. *Australasian Journal of Combinatorics* 54:19–36.

Parker, D.B., R.F. Westhoff, and M.J. Wolf. 2006. Two-path convexity in clonefree regular multipartite tournaments. *Australasian Journal of Combinatorics* 36:177–196.

Parker, D.B., R.F. Westhoff, and M.J. Wolf. 2008. On two-path convexity in multipartite tournaments. *European Journal of Combinatorics* 29:641–651.

Parker, D.B., R.F. Westhoff, and M.J. Wolf. 2009. Convex independence and the structure of clone-free multipartite tournaments. *Discussiones Mathematicae Graph Theory* 29(1):51–69.

Parvathy, K.S. and A. Vijayakumar. 1998. Geodesic iteration number. In *Conference on graph connections*, 91–94. New Delhi.

Patkós, B. 2020. On the general position problem on Kneser graphs. *Ars Mathematica Contemporanea* 18:273–280.

Pelayo, I.M. 2013. *Geodesic convexity in graphs*, viii+112. Berlin: Springer.

Peleg, D. 1998. Size bounds for dynamic monopolies. *Theoretical Computer Science* 86(2):263–273.

Penso, L., F. Protti, D. Rautenbach, and U.S. Souza. 2015. Complexity analysis of P3-convexity problems on bounded-degree and planar graphs. *Theoretical Computer Science* 607:83–95.

Polat, N. 1995. A Helly theorem for geodesic convexity in strongly dismantlable graphs. *Discrete Mathematics* 140(1–3):119–127.

Polat, N. 2000. On infinite bridged graphs and strongly dismantlable graphs. *Discrete Mathematics* 211(1–3):153–166.

Polat, N. 2003. On constructible graphs, locally Helly graphs, and convexity. *Journal of Graph Theory* 43(4):280–298.

Protti, F., and J.V.C. Thompson. 2023. All-path convexity: Combinatorial and complexity aspects.

Radon, J. 1921. Mengen konvexer Kürper, die einen gemeinsamen Punkt enthalten. *Mathematische Annalen* 83(1):113–115.

Ramos, I., V.F. dos Santos, and J.L. Szwarcfiter. 2014. Complexity aspects of the computation of the rank of a graph. *Discrete Mathematics & Theoretical Computer Science* 16(2) (cit. on pp. 27, 52).

Roberts, F.S. 1969. In difference graphs. In *Proof techniques in graph theory*, ed. by F. Harary, 139–146. New York: Academic Press.

Robertson, N., and P.D. Seymour. 1986. Graph minors. II. Algorithmic aspects of tree-width. *Journal of Algorithms* 7(3):309–322.

Sampathkumar, E. 1984. Convex sets in a graph. *Indian Journal of Pure and Applied Mathematics* 15:1065–1071.

Schaefer, T.J. 1978. On the complexity of some two-person perfect-information games. *Journal of Computer and System Sciences* 16(2):185–225.

Sierksma, G. 1975. Carathéodory and Helly-numbers of convex-product structures. *Pacific Journal of Mathematics* 61(1):275–282.

Sierksma, G. 1976. Axiomatic Convexity Theory and the Convex Product Space. PhD Thesis. University of Groningen.

Sierksma, G. 1977. Relationships between Carathéodory, Helly, Radon and Exchange numbers of convexity spaces. *Nieuw Archief voor Wiskunde* 3(25):115–132 (cit. on p. 29).

Sierksma, G., H.M. Mulder, and M. Changat. 2000. *Convexities related to path properties on graphs; a unified approach*. Research Report 00A45. University of Groningen.

Steinitz, E. 1916. Bedingt konvergente Reihen und konvexe Systeme (Schlus). *J. für die Reine und Angewandte Mathematik* 146:1–52.

Stockmeyer, L.J., and A.R. Meyer. 1973. Word problems requiring exponential time (preliminary report). In *Proceedings of the Fifth Annual ACM Symposium on Theory of Computing – ACM STOC'73*, 1–9. New York: ACM .

Szekeres, G., and L. Peters. 2006. Computer solution to the 17-point Erdős-Szekeres problem. *The ANZIAM Journal* 48(2):151–164.

Thomas, E.J., and U. Chandran S.V. 2020. Characterization of classes of graphs with large general position number. *AKCE International Journal of Graphs and Combinatorics* 17(3):935–939.

Tian, J., K. Xu, and S. Klavžar. 2021. The general position number of the Cartesian product of two trees. *Bulletin of the Australian Mathematical Society* 104:1–10.

Tian, J., and K. Xu. 2021. The general position number of Cartesian products involving a factor with small diameter. *Applied Mathematics and Computation* 403:126206.

Tverberg, H. 1966. A generalization of Radon's theorem. *Journal of the London Mathematical Society* 41(1):123–128.

Ullas Chandran, S.V., S. Klavžar, N.P.K., and R. Sampaio. 2024. The general position avoidance game and hardness of general position games. *Theoretical Computer Science* 988:114370.

Van de Vel, M.L.J. 1993. *Theory of convex structures*. Vol. 50, xv+540. Amsterdam: Elsevier (cit. on pp. 3, 6, 8, 11, 13, 23–25, 29, 61).

Yannakakis, M. 1981. Node-deletion problems on bipartite graphs. *SIAM Journal on Computing* 10(2):310–327.

Yao, Y., M. He, and S. Ji. 2022. On the general position number of two classes of graphs. *Open Mathematics* 20(1):1021–1029.

Zermelo, E. 1913. Über eine Anwendung der Mengenlehre auf die Theorie des Schachspiels. In *Proceedings of the 5th International Congress of Mathematics*, 501–504.

Zuckerman, D. 2006. Linear degree extractors and the inapproximability of max clique and chromatic number. In *STOC'06*, 681–690. New York: ACM.

Index of Notations

J. Araújo et al., *Introduction to Graph Convexity*, SBMAC Collection on Applied and Computational Mathematics, https://doi.org/10.1007/978-3-031-84128-6

Index of Authors

J. Araújo et al., *Introduction to Graph Convexity*, SBMAC Collection on Applied and Computational Mathematics, https://doi.org/10.1007/978-3-031-84128-6

Remissive Index

J. Araújo et al., *Introduction to Graph Convexity*, SBMAC Collection on Applied and Computational Mathematics, https://doi.org/10.1007/978-3-031-84128-6

MIX
Papier aus verantwortungsvollen Quellen
Paper from responsible sources
FSC® C105338

If you have any concerns about our products,
you can contact us on
ProductSafety@springernature.com

In case Publisher is established outside the EU,
the EU authorized representative is:
Springer Nature Customer Service Center GmbH
Europaplatz 3, 69115 Heidelberg, Germany

Printed by Libri Plureos GmbH
in Hamburg, Germany